SpringerBriefs in Agriculture

More information about this series at http://www.springer.com/series/10183

Samiha A.H. Ouda • Abd El-Hafeez Zohry
Huda Alkitkat • Mostafa Morsy
Tarek Sayad • Ahmed Kamel

Future of Food Gaps in Egypt

Obstacles and Opportunities

 Springer

Samiha A.H. Ouda
Water Requirements and Field Irrigation
 Research Department, Soils, Water
 and Environment Research Institute
Agricultural Research Center
Giza, Egypt

Huda Alkitkat
Fellow of Academy of Scientific Research
 and Technology
Cairo, Egypt

Tarek Sayad
Department of Astronomy and Meteorology,
 Faculty of Science
Al Azhar University
Cairo, Egypt

Abd El-Hafeez Zohry
Crops Intensifications Research
 Department, Field Crops Research
 Institute
Crops Agricultural Research Center
Giza, Egypt

Mostafa Morsy
Department of Astronomy and Meteorology,
 Faculty of Science
Al Azhar University
Cairo, Egypt

Ahmed Kamel
Crops Intensifications Research
 Department, Field Crops Research
 Institute
Crops Agricultural Research Center
Giza, Egypt

ISSN 2211-808X   ISSN 2211-8098   (electronic)
SpringerBriefs in Agriculture
ISBN 978-3-319-46941-6   ISBN 978-3-319-46942-3   (eBook)
DOI 10.1007/978-3-319-46942-3

Library of Congress Control Number: 2016960754

Printed on acid-free paper

This Springer imprint is published by Springer Nature
The registered company is Springer International Publishing AG
The registered company address is: Gewerbestrasse 11, 6330 Cham, Switzerland

Contents

Chapter 1
Introduction

Samiha A.H. Ouda and Abd El-Hafeez Zohry

In accordance with the United Nations' agenda for the 2030 for sustainable development, namely the second goal: "End hunger, achieve food security, improve nutrition and promote sustainable agriculture", the relationship between food security and population growth is a very critical and hot issue that has been raised in most of the developing countries that have overpopulation problem like Egypt. Egypt population size increased rapidly during the last three decades from 48 million in 1986 to about 90 million in 2016.

In 1960, Egypt was self-sufficient in almost all basic food commodities, with the exception of wheat, of which the country had a self-sufficiency ratio (domestic production in relation to consumption) of 70 %. The self-sufficiency ratio declined dramatically for most products during the 1970s and 1980s, and economists began to speak of a serious food gap (potential biological-genetic yield versus actual farm yield) in Egypt (Metz 1990). Currently, there is a gap between production and consumption of cereal crops, oil crops, sugar crops, legume crops and forage crops. Reducing yield gaps is one of the main goals of food security research in Egypt.

Land resources in Egypt are abundant, because we live on only 4 % of Egypt's area. However, our water resources became limited. Egypt has reached a state, where the quantity of available water is imposing limits on its national economic development, where Egypt has passed the scarcity threshold (Ministry of Irrigation

S.A.H. Ouda (✉)
Water Requirements and Field Irrigation Research Department, Soils, Water and Environment Research Institute, Agricultural Research Center, Giza, Egypt
e-mail: samihaouda@yahoo.com

A.E.-H. Zohry
Crops Intensifications Research Department, Field Crops Research Institute,
Crops Agricultural Research Center, Giza, Egypt
e-mail: abdelhafeezzohry@yahoo.com

© Springer International Publishing AG 2017
S. Ouda et al., *Future of Food Gaps in Egypt*, SpringerBriefs in Agriculture,
DOI 10.1007/978-3-319-46942-3_1

and Water Resources 2014). Nevertheless, water management on field level is poor, with low application efficiency, which endures large water losses to the ground water.

Thus, scarcity of water will put more pressure on water resources distribution between economic sectors, especially agriculture, where food is produced.

To solve food gaps problems in Egypt, unconventional procedures are needed to increase crops productivity, manage irrigation water more efficiently and increase crops production in short time. Since surface irrigation is prevailing in Egypt, it is important to test options to increase its low application efficiency, i.e. 60 %. Managing irrigation water more efficiently can be done through changing cultivation methods from cultivation in basins or on narrow furrows to raised beds. Raised beds cultivation can increase crops productivity by 15 % and save on the applied water by 20 %. Other benefits can be obtained from raised beds cultivation, i.e. reduction in the applied fertilizer amount, which reduce ground water pollution and increase farmer's net revenues (Aboelenein et al. 2011).

Intercropping systems is another procedure can be done to help in solving food insecurity problem. Intercropping provides year-round ground cover, or at least for a longer period than monocultures, in order to protect the soil from desiccation and erosion. By growing more than one crop at a time in the same field, farmers maximize water use efficiency, maintain soil fertility, and minimize soil erosion, which are the serious drawbacks of mono-cropping (Hoshikawa 1991). Furthermore, intercropping increase land productivity and it also can save a sum of irrigation water, thus increase water productivity (Kamel et al. 2016). Increased crop production often observed in intercrops compared to sole crops has been attributed to enhanced resource use (Szumigalski and Van-Acker 2008). Sivakumar (1993) also reported that efficient and complete use of growth resources, such as solar energy, soil nutrients and water is one of the advantages of intercropping systems over sole crops.

Climate change is one of the overwhelming environmental threats that are defined as a long-term alteration in the global weather patterns, including temperature, precipitation, soil moisture, sea level, and storm activity. Such climate change is a potential consequence for releasing the greenhouse gases (GHGs) that accumulated in the atmosphere, resulting in global warming (El Massah and Omran 2014). Furthermore, climate change is expected to increase potential evapotranspiration due to higher temperature, solar radiation and wind speed (Abtew and Melesse 2013), which will affect the hydrological system and water resources (Shahid 2011). Thus, agriculture will highly suffer from these expected effects of climate change. Unfortunately, there is high uncertainty level associated with the projection of climate change scenarios (IPCC 2007). However, Morsy (2015) and Morsy et al. (2016) were successful to used statistical analysis procedures to increase the certainty in the projection of the output of global climate models (AR5). Thus, they were able to increase the accuracy of the projection of climate change effect on crops production.

In this book, we are concern about current food gaps in Egypt, namely the current gap between production and consumption of wheat, maize and faba bean. All these crops are important components of Egyptian food basket. As a result of high population growth rate, these food gaps are in continuous increase and it is expected to increase in the future under water scarcity induced by climate change. Thus, quantification of the effect of different suggested management practices are very important in determining its contribution in reducing current and future food gaps. Furthermore, these practices can be used as adaptation strategies under climate change to reduce vulnerability of the selected crops to climate change and reduce its production-consumption gap in the future. These adaptation strategies can meet the increasing demand for food in 2030 due to rapid population growth, water scarcity and climate change risks.

In this analysis, we studied the interrelation among population (consumer of wheat, maize and faba bean), land and water resources, as well as weather resources to suggest solution to reduce the gap between production of these crops and its consumption. To do so, Egypt was divided into three regions. The first region is Lower Egypt included all Nile Delta governorates and Sinai. Middle Egypt included all Middle Egypt governorates, in addition to Marsa Matrouh governorate. Upper Egypt included all Upper Egypt governorates, in addition to El-Wadi El-Gedid and Red Sea governorates (Fig. 1.1).

Data on the cultivated area of the selected crops and productivity in 2012/2013 for wheat and faba bean and other studied winter crops used in the study were collected from publish data by Ministry of Agriculture and Land Reclamation in Egypt.

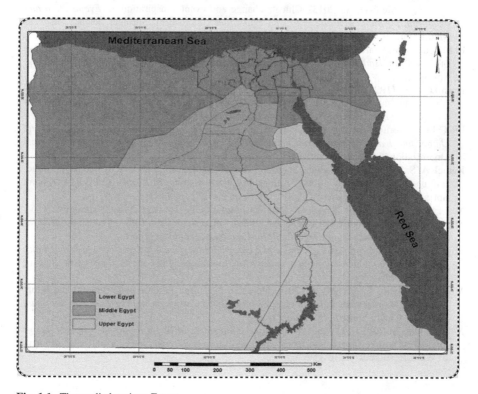

Fig. 1.1 The studied regions Egypt

Similar procedure was done for maize data and other studied summer crops in 2013. Weather data for the studied growing seasons were collected. Water requirements for all the selected crops were calculated using BISm model (Snyder et al. 2004) in 2012 and 2013, as well as under climate change in 2029 and 2030.

Several assumptions were used to implement the assessment of the impact of climate change on the selected crops. We assumed that the productivity of the selected crops will not be negatively affected by the abiotic stress of climate change, i.e. heat and water stresses. This assumption was supported by the new released cultivars of the selected crops by the Egyptian Agricultural Research Center. These cultivars are high yielding with high tolerance to heat and water stresses. We also assumed that the Egyptian share of the River Nile water will not affected by climate change. This assumption relayed on previous results obtained by Nour El-Din (2013) and stated that the results of some general circulation models (GCMs) are in the direction of an increase in precipitation in East Africa.

References

Aboelenein, R., Sherif, M., Karrou, M., Oweis, T., Benli, B., & Farahani, H. (2011). Towards sustainable and improved water productivity in the old lands of Nile Delta. In *Water benchmarks of CWANA – Improving water and land productivities in irrigated systems.*

Abtew, W., & Melesse, A. (2013). Climate change and evapotranspiration. In *Evaporation and evapotranspiration: Measurements and estimations.* Dordrecht: Springer Science Business Media. doi: 10.1007/978-94-007-4737-113.

El Massah, S., & Omran, G. (2014). Would climate change affect the imports of cereals? The case of Egypt. In *Handbook of climate change adaptation* (pp. 657–685). Berlin/Heidelberg: Springer.

Hoshikawa, K. (1991). Significance of legumes crops in intercropping, the productivity and stability of cropping system. In C. Johanson, K. K. Lee, & K. L. Saharawat (Eds.), *Phosphorus nutrition of grain legume in the semi arid tropics* (pp. 173–176). Patachcheru: ICRISAT.

IPCC Intergovernmental Panel on Climate Change. (2007). *Intergovernmental panel on climate change fourth assessment report: Climate change 2007.* Synthesis report.World Meteorological Organization, Geneva, Switzerland.

Kamel, A. S., Zohry, A. A., & Ouda, S. (2016). Unconventional solution to increase crops production under water scarcity. In *Major crops and water scarcity in Egypt* (pp. 99–114). Cham: Springer Publishing House.

Metz, H. C. (1990). *Egypt: A country study.* Washington: GPO for the Library of Congress.

Ministry of Irrigation and Water Resources. (2014). *Water Scarcity in Egypt: The urgent need for regional cooperation among the Nile Basin Countries.* Technical report.

Morsy, M. (2015). *Use of regional climate and crop simulation models to predict wheat and maize productivity and their adaptation under climate change.* PhD thesis. Faculty of Science Al-Azhar University.

Morsy, M., Sayad, T., & Ouda, S. (2016). Potential evapotranspiration under present and future climate. In *Management of climate induced drought and water scarcity in Egypt: Unconventional solutions.* Cham: Springer Publishing House.

Nour El-Din, M. (2013). *Climate change risk management in Egypt proposed: Climate change adaptation strategy for the ministry of water resources and irrigation in Egypt.* Cairo: Ministry of Water Resources and Irrigation.

Shahid, S. (2011). Impacts of climate change on irrigation water demand in Northwestern Bangladesh. *Climatic Change, 105*(3–4), 433–453.

Sivakumar, M. V. K. (1993). Growth and yield of millet and cowpea in rely and intercrop systems in the Sahelian Zones in years when the onset of the rainy season is early. *Experimental Agriculture, 29*(4), 417–427.

Snyder, R. L., Orang, M., Bali, K., & Eching, S. (2004). Basic irrigation scheduling (BIS). http://www.waterplan.water.ca.gov/landwateruse/wateruse/Ag/CUP/Californi/Climate_Data_010804.xls

Szumigalski, A. R., & Van-Acker, R. C. (2008). Intercropping: Land equivalent ratios, light interception, and water use in annual intercrops in the presence or absence of in-crop herbicides. *American Agronomy Journal, 100*, 1145–1154.

Chapter 2
Present and Future Water Requirements for Crops

Mostafa Morsy, Tarek Sayad, and Samiha A.H. Ouda

Introduction

Egypt is located in the North-Eastern corner of the African continent and lies between 22° to about 32° N and 24° to about 36° E, with a total area of about one million square kilometers. It is bordered on the west by Libya, on the north by the Mediterranean Sea, on the south by Sudan, and on the east by the Gaza Strip and the Red Sea. The inhabited area of the country is confined to the narrow strip of the Nile valley, from Aswan in the south to Cairo in the north. Furthermore, the Nile Delta covers the area from Cairo to the shoreline of the Mediterranean Sea, between the cities of Damietta in the east and Rashid in the west.

The climate of Egypt is characterized by hot dry summers and mild winters prevail with relatively low, irregular, and unpredictable rainfall. The average daily temperature ranges from 17 to 20 °C along the Mediterranean to more than 25 °C in Upper Egypt along the Nile (EEAA 2010). Figure 2.1 displays average annual temperatures across Egypt.

Precipitation is generally very low. It is highest along the Mediterranean, where it average to more than 200 mm/year. Precipitation rates drop quickly as one moves away from the coast. It average of 20 mm/year in Middle Egypt to 2 mm/year in Upper Egypt (Fig. 2.2). Thus, most of Egypt is a desert and can be classified as arid. The exception is the slightly wetter Mediterranean coast, which can be considered

M. Morsy (✉) • T. Sayad
Department of Astronomy and Meteorology, Faculty of Science, Al Azhar University,
Cairo, Egypt
e-mail: mustafa_meteorology@yahoo.com; ta_sayad@yahoo.com

S.A.H. Ouda
Water Requirements and Field Irrigation Research Department, Soils, Water and Environment
Research Institute, Agricultural Research Center, Giza, Egypt
e-mail: samihaouda@yahoo.com

© Springer International Publishing AG 2017
S. Ouda et al., *Future of Food Gaps in Egypt*, SpringerBriefs in Agriculture,
DOI 10.1007/978-3-319-46942-3_2

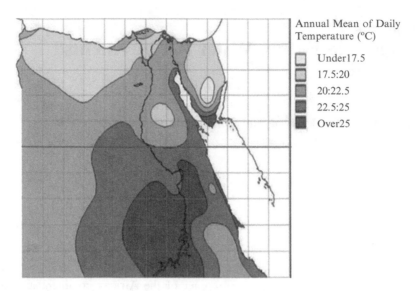

Fig. 2.1 Average annual temperatures (°C) in Egypt (Source: EEAA 2010)

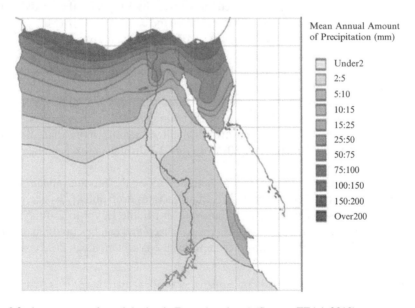

Fig. 2.2 Average annual precipitation in Egypt (mm/year) (Source: EEAA 2010)

semi-arid. Generally, the small amount of rains that fall comes in the winter, and hence Egypt has a Mediterranean climate.

Climate plays an important role in crop production. Crops growth periods, crops water requirements and scheduling irrigation for crops are dependent on weather conditions. The relationships between climate, crop, water and soil are complex with many biological processes involved (Rao et al. 2011).Therefore, it is important to study weather pattern in a region to best identify its effect of the growing crops and irrigation water needs. Thus, the objectives of this chapter were: (i) to describe annual and seasonal climate of Egypt with respect to maximum and minimum temperatures, wind speed and solar radiation; (ii) to calculate water requirements for wheat, maize and faba bean grown in three governorates to represent the three main regions in Egypt in the growing season of 2012/2013; (iii) to calculate water requirements for the above crops in the growing season of 2029/2030 using more accurate AR5 scenarios.

Egypt's Seasonal Climate

There are four distinguished seasons in Egypt: winter (from December to February), spring (from March to May), summer (June to August) and autumn (from September to November). During the winter season, Egypt is under the influence of mid-latitude weather regime, which is characterized by the frequent passage of upper westerly troughs associated with surface depressions. The monthly average of these depressions is from three to five. These depressions cause north-easterly winds with relatively stable conditions west of the depression that affect the Egyptian regions. During the passage of these depressions along the east Mediterranean, thick layers of low and medium clouds cover the northern part of Egypt and sometimes extend to the south. These cloud clusters are usually associated with torrential rain and in many cases with thunder storms (El-Fandy 1948).

Within the spring season, the main climatic feature is the southward shift of the tracks of depressions. The centers of the depressions move either along the coast line of north Africa or farther south, where they are known as desert or Khamasen depressions. The average frequency of these latter depressions is three to four per month. After the passage of the depression, northerly winds prevail over Egypt with possible strong guests that cause rising sand (dust) in open areas especially in Upper Egypt. These depressions, when associated with strong sub-tropical jet stream in the upper troposphere, generally produce strong surface winds that cause large scale rising sand over most Egyptian areas. In such cases large scale sand storm is possible (El-Fandy 1948).

During the summer season, the main forcing for this season is the westward extension of the Indian monsoon low pressure. This is a thermal shallow low acting over south west Asia, and provides Egypt with hot and humid north easterly surface winds. Occasionally, during this season, the Indian monsoon low pressure is weakened and shrinks towards the east paving the way to the sub-tropical high pressure. In this case the subtropical high pressure, which is located over west Europe, extends to the east to cover the east Mediterranean Sea. As a consequence, a rather mild and dry north westerly wind invades the northern parts of Egypt (Hasanean and Abdel Basset 2006).

During the autumn season, Egypt is under the influence of northward extension of Sudan monsoon low along Red Sea with southward extension of upper air trough that cause heavy rain over the eastern part of Egypt (Hasanean and Abdel Basset 2006).

Seasonal Patterns of Meteorological Parameters in Egypt

The distinctive climate features of each season for the three studied regions of Egypt are very important in the calculation of water requirements for crops. Therefore, long term monthly means of climatic elements (maximum and minimum temperatures, wind speed and solar radiation) at the earth's surface during the climate period from 1981 to 2010 were collected. These data were derived from NCEP/NCAR Reanalysis data with horizontal resolution of 2.5° latitude × 2.5° longitude global grids (144×73 grid points) from 0.0° to 357.5° E and from 90.0° to 90.0 °S (http://www.esrl.noaa.gov/psd/data/gridded/data.ncep.reanalysis.derived.html).

Maximum Temperature

The maximum temperature in Egypt increases gradually southward in all seasons following the apparent position of the sun. The minimum value of maximum temperature occurs in the winter, while it tends to reach its maximum value in the summer. One can notice that gradient of maximum temperature intensifies during summer and spring seasons over northern part of Egypt and decline over the rest of Egypt (Fig. 2.3). Maximum temperature during winter tends to reach its maximum value of 18 °C over the northern part and maximum value of 24 °C over the southern part with small gradient of temperature (Fig. 2.3a). Maximum temperature during spring and autumn seasons ranges from 21 °C in the north to 36 °C in the south (Fig. 2.3b and 2.3d). Whereas, maximum temperature during summer season ranges from 30 to 42 °C with maximum core over Upper Egypt (Fig. 2.3c).

Minimum Temperature

Most of Middle Egypt has the lowest values of minimum temperature in all seasons. Where, the lowest value of minimum temperature occurs in the winter, while it tends to be in its maximum value in the summer. The gradient of minimum temperature intensifies during winter and autumn seasons over northern part of Egypt and decline over the rest of Egypt (Fig. 2.4). Minimum temperature during winter tends to low as 3 °C over most of Egypt, except the northern part and south eastern part (Fig. 2.4a). Minimum temperature during spring season is around 12 °C covers most of

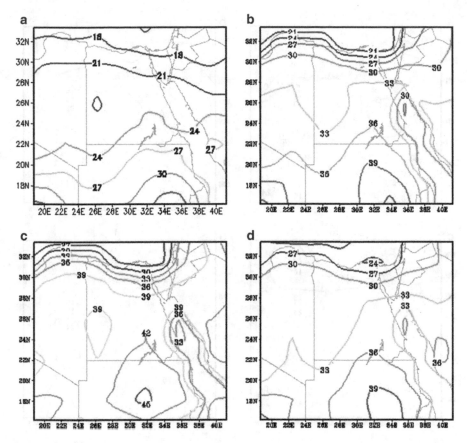

Fig. 2.3 Seasonal average of maximum temperature (°C) during the period of 1981–2010. (**a**) Winter Tmax (°C). (**b**) Spring Tmax (°C). (**c**) Summer Tmax (°C). (**d**) Autumn Tmax (°C)

Egypt (Fig. 2.4b). Furthermore, 21 °C minimum temperature during summer season is the prevailing temperature over all Egypt except latitudinal band from 28 to 30 °N (Fig. 2.4c). Finally, 15 °C minimum temperature during autumn season is the dominant temperature over all Egypt, except north and south east parts (Fig. 2.4d).

Wind Speed

Figure 2.5 illustrates 10 m wind speed in the four seasons in Egypt. Wind speed increases gradually from minimum value of 1.5 m/s in winter to maximum value of 4 m/s in summer. The maximum core of wind speed is found over south west of Egypt, while minimum core is found over Middle Egypt in all seasons. The strong gradient of wind over northern part is according to Mediterranean depression in winter season.

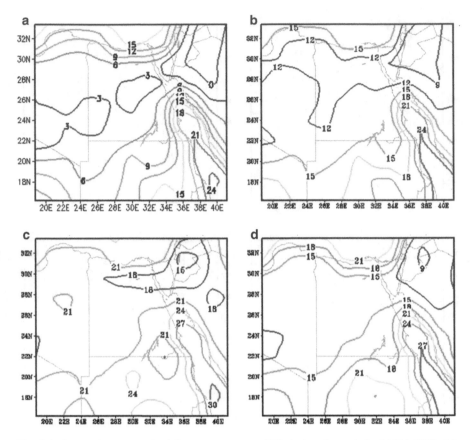

Fig. 2.4 Seasonal average of minimum temperature (°C) during the period of 1981–2010. (**a**) Winter Tmin (°C). (**b**) Spring Tmin (°C). (**c**) Summer Tmin (°C). (**d**) Autumn Tmin (°C)

Solar Radiation

Figure 2.6 indicated that downward short wave radiation decreases gradually from south to north according to the apparent position of sun and reaches its maximum value in the summer season. Downward short wave radiation has a strong gradient in winter, while it has a weak gradient in summer. The dominant value of downward short wave radiation is 32 MJ/m² during summer in most Egypt. The difference between north and south values of downward short wave radiation is large in winter. In addition, there is a small variability in downward short wave radiation in both transient seasons.

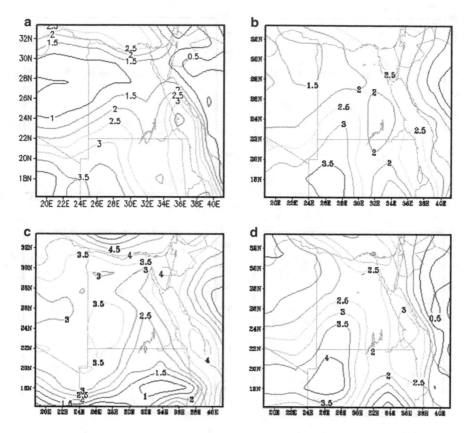

Fig. 2.5 Seasonal average of wind speed (m/s) during the period of 1981–2010. (**a**) Winter Wind Speed (m/s). (**b**) Spring Wind Speed (m/s). (**c**) Summer Wind Speed (m/s). (**d**) Autumn Wind Speed (m/s)

The Selected Governorates

Three governorates were selected to represent the climate of Lower, Middle and Upper Egypt, namely El-Monofia, El-Fayom and Qena governorates, respectively (Table 2.1).

Solar radiation, maximum and minimum temperatures, and wind speed are considered the most important climatic parameters that describe the main features for each weather station and consequently essential for the calculation of water requirements for the cultivated crops. For each governorate, the values for these parameters averaged from 1984 to 2015 were obtained. Regarding to El-Monofia governorate, Fig. 2.7a showed that solar radiation attains its maximum in June and decreases gradually forward and backward to December and January, respectively. There is 1 month lag between maximum and minimum solar radiation value and their corresponding values of maximum temperature (Fig. 2.7b). The same behavior can be

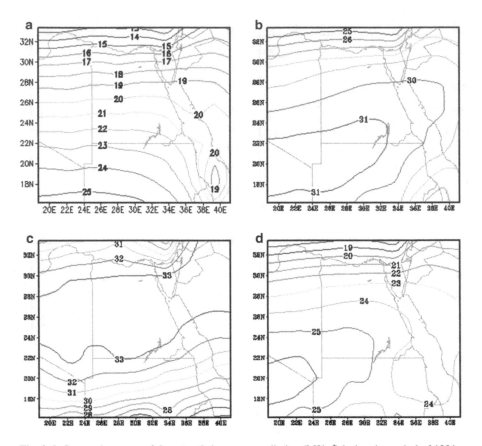

Fig. 2.6 Seasonal average of downward short wave radiation (MJ/m²)during the period of 1981–2010. (**a**) Winter Radiation (MJ/m²). (**b**) Spring Radiation (MJ/m²). (**c**) Summer Radiation (MJ/m²). (**d**) Autumn Radiation (MJ/m²)

noticed for minimum temperature but with 2 months lag (Fig. 2.7c). Whereas, wind speed has its maximum value between April and June, while it reaches its minimum value in December (Fig. 2.7d).

The same pattern can be observed in El-Fayom (Fig. 2.8) and in Qena (Fig. 2.9), except for wind speed, where its maximum is found in June only (Figs. 2.8d and 2.9d).

Water Requirements for the Studied Crops

The term crop water requirement is defined as the amount of water required to compensate the evapotranspiration loss from the cropped field (USDA, 1993). ICID-CIID (2000) describes it as the total water needed for evapotranspiration, from

Table 2.1 Latitude, longitude and elevation above sea level for the selected Governorates

Governorate	Latitude (°N)	Longitude (°E)	Elevation above sea level (m)
Lower Egypt	30.36	31.01	17.9
El-Monofia			
Middle Egypt	29.18	30.51	30.0
El-Fayom			
Upper Egypt	26.10	32.43	72.6
Qena			

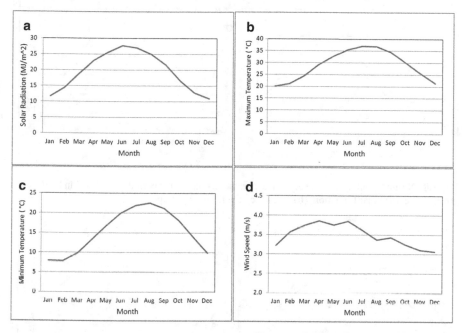

Fig. 2.7 Monthly averages of normal climate parameters during the period 1984–2015 in El-Monofia governorate. (**a**) Solar Radiation. (**b**) Tmax. (**c**) Tmin. (**d**) Wind Speed

planting to harvest for a given crop in a specific climate, when adequate soil water is maintained by rainfall and/or irrigation so that it does not limit plant growth and crop yield. Although the values of crop evapotranspiration (ETc) and crop water requirement are identical, crop water requirement refers to the amount of water that needs to be supplied, while ETc refers to the amount of water that is lost through evapotranspiration (Allen et al. 1998).

The BISm model (Snyder et al. 2004) was used to calculate monthly reference evapotranspiration (ETo) and water requirements for the selected crops. The model calculates ETo using Penman-Monteith equation (Monteith 1965) as presented in the United Nations FAO Irrigation and Drainage Paper (FAO 56) by Allen et al. (1989). Weather data for 2012/2013 season was used in this analysis because it is coincided with the data obtained for cultivated areas and production of the selected crops. The latitude and elevation of the weather station for the three studied gover-

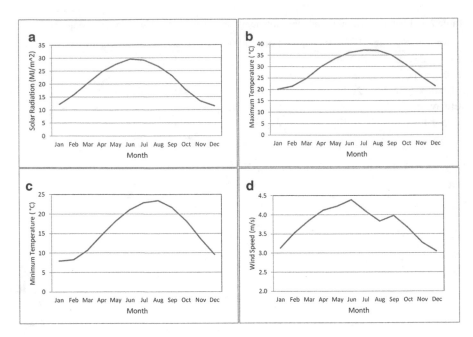

Fig. 2.8 Normal monthly average of climate parameters during the period 1984–2015 in El-Fayom governorate. (**a**) Solar Radiation. (**b**) Tmax. (**c**) Tmin. (**d**) Wind Speed

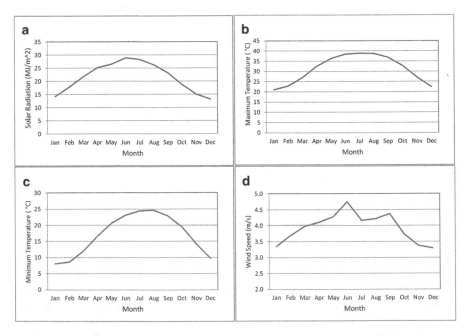

Fig. 2.9 Normal monthly average of climate parameters during the period 1984–2015 in Qena governorate. (**a**) Solar Radiation. (**b**) Tmax. (**c**) Tmin. (**d**) Wind Speed

Table 2.2 Water requirements (WR, m³/ha) for wheat under different irrigation systems in 2012/2013

	WR under surface irrigation	WR under raised beds	WR under sprinkler system
Lower Egypt	7122	5698	5342
Middle Egypt	7367	5893	5525
Upper Egypt	8200	6560	6150
Average	7563	6050	5672

norates were required as input. After calculating daily means of weather data by month, the model uses a cubic spline curve fitting subroutine to estimate daily ETo rates for the entire year. To calculate water requirements for the selected crops at each governorate, planting and harvest dates for each crop were input. The model calculated crop coefficient and accounted for water depletion from root zone in the growing season of each selected crop. Water requirements were calculated for wheat (Table 2.2) irrigated with surface irrigation (60 % application efficiency), under cultivation on raised beds (20 % lower than applied amount under surface irrigation) and under irrigation with sprinkler system (80 % application efficiency).

Water requirements for faba bean (Table 2.3) and maize (Table 2.4) irrigated with surface irrigation, raised beds cultivation and irrigated with drip system (90 % application efficiency) were also calculated.

Projected Water Requirements Under Climate Change

Climate change is expected to have many adverse impacts on various sectors, yet agriculture is considered to be the most tangible affected sector, as any alteration in the prevailing temperature or precipitation patterns will disturb the agricultural sector as a whole, including crop yields, crops water requirements and soil fertility (El-Massah and Omran 2014). Abtew and Melesse (2013) indicated that climate change will increase temperature, solar radiation and wind speed. These three weather elements are the main contributors in ETo and it will consequently affect the hydrological system and water resources (Shahid 2011).

The SRES scenarios of IPCC Fourth Assessment Report (AR4) released in 2007 by the Intergovernmental Panel on Climate Change was the main source for climate change scenarios to be used in studying climate change effect on different aspects (Wayne 2013). Furthermore, the Fifth Assessment Report (AR5) of the Intergovernmental Panel on Climate Change (IPCC 2013) is the recent that contained a large number of comprehensive climate models and Earth System Models, whose results form the core of the climate system projections.

In Egypt, Khalil (2013) used SERS scenarios to calculate ETo values in the future and indicated that it will increase under climate change compared to current climate. Furthermore, using AR4 scenarios, Ouda et al. (2016) indicated that ETo values will increase depending on the location of the region, where it will be lower

Table 2.3 Water requirements (WR, m³/ha) for faba bean under different irrigation systems in 2012/2013

	WR under surface irrigation	WR under raised beds	WR under drip system
Lower Egypt	6512	5210	4341
Middle Egypt	6760	5408	4507
Upper Egypt	8822	7058	5882
Average	7365	5892	4910

Table 2.4 Water requirements (WR, m³/ha) for maize under different irrigation systems in 2012/2013

	WR under surface irrigation	WR under raised beds	WR under drip system
Lower Egypt	10,829	8663	7219
Middle Egypt	11,807	9445	7871
Upper Egypt	12,967	10,373	8644
Average	11,867	9494	7912

in North Egypt, compare to the Middle and South of Egypt. They also indicated that water requirements for wheat will increase by 9 % in the Nile Delta and 18 % in both Middle and Upper Egypt in 2040. Similarly, water requirements for maize are expected to increase by 13 % in the Nile Delta, 16 % in Middle Egypt and 17 % in Upper Egypt. Using AR5 scenarios, water requirements for faba bean is expected to increase by 3 % in the Nile Delta (Zohry and Ouda 2016a) and by 17 % in Upper Egypt in 2030 (Zohry and Ouda 2016b).

Projected Weather Data Under Climate Change

The recent report by IPCC in 2013 contained a large number of comprehensive climate models produce new Representative Concentration Pathways (RCP2.6, RCP4.5, RCP6.0 and RCP8.5), which based on Coupled Model Inter-comparison Project Phase 5 (CMIP5) to replace AR4 scenarios for more accurate assessment of the effect of climate change (Solomon et al. 2007). However, there is a level of uncertainty associated with the projections of climate change in the future using the developed models by IPCC. To lower that uncertainty, Morsy (2015) developed a methodology to determine the most suitable global climate model and scenarios for four governorates (Kafr El-Sheik, El-Gharbia, El-Minia and Sohag) in Egypt. This methodology depended on performing comparison between measured meteorological data and the projected data from four global climate models represented by its four RCPs scenarios during the period from 2006 to 2012. His results indicated that

Table 2.5   List of selected CMIP5 GCM models and their horizontal resolutions

Model	Institution	Horizontal resolution
BCC-CSM1-1	Beijing Climate Center, China Meteorological Administration, China.	2.80° × 2.80°
CCSM4	National Centre for Atmospheric Research (NCAR), Community Climate System Model, USA.	1.25° × 0.94°
GFDL-ESM2G	National oceanic and Atmospheric Administration (NOAA), Geophysical Fluid Dynamics Laboratory (GFDL), USA.	2.02° × 2.00°
MIROC5	Atmosphere and ocean Research Institute (The University of Tokyo), National Institute for Environmental Studies, and Japan Agency for Marine-Earth Science and Technology, Japan.	1.40° × 1.40°

RCP6.0 scenario from CCSM4 model was more accurate to project the meteorological variables because it achieved the closest values of goodness of fit test between measured and projected meteorological data. This methodology was applied here for the three studied governorates.

In this methodology, four models were selected, where it have differing levels of sensitivity to Green House Gases (GHG) forcing and focused on daily RCPs climate change scenarios. The selected models are listed in Table 2.5.

Each model produces four RCPs scenarios to represent a larger set of mitigation scenarios and have different targets in terms of radiative forcing in 2100. Each RCP defines a specific emissions trajectory. The emission pathway of RCP2.6 is representative of scenarios in the literature that lead to very low greenhouse gas concentration levels. It is a "peak-and-decline" scenario, where its radiative forcing level first reaches a value of around 3.1 W/m² by mid-century, and returns to 2.6 W/m² by 2100. In order to reach such radiative forcing levels, greenhouse gas emissions and indirectly emissions of air pollutants are reduced substantially over time (Van Vuuren et al. 2006, 2007). Regarding to RCP4.5, it is a stabilization scenario, in which total radiative forcing is stabilized shortly after 2100, without overshooting the long run radiative forcing target level (Smith and Wigley 2006; Clarke et al. 2007; Wise et al. 2009). Furthermore, RCP6.0 is a stabilization scenario, in which total radiative forcing is stabilized shortly after 2100, without overshoot, by the application of a range of technologies and strategies for reducing greenhouse gas emissions (Fujino et al. 2006; Hijioka et al. 2008). With respect to RCP8.5, it is characterized by increasing greenhouse gas emissions over time, where it is representative of scenarios in the literature that lead to high greenhouse gas concentration levels (Riahi et al. 2007).

Goodness of fit test between the measured and projected meteorological data (solar radiation, maximum, minimum temperatures and wind speed) by the RCPs for selected governorates was done using the following tests:

Willmott Index of Agreement (d-stat)

It is the standardized measure of the degree of model prediction error which varies between 0 and 1. A value of 1 indicates a perfect match, and value of 0 indicates no agreement at all (Willmott 1981).

$$d-stat = 1 - \frac{\sum\limits_{n}^{i=1}(O_i - S_i)^2}{\sum\limits_{n}^{i=1}[(|(S_i - \bar{O})| + |(O_i - \bar{O})|)^2]} \tag{2.1}$$

Where Oi, Ō and Si represent the observed, observed average and simulated values.

Coefficient of Determination (R2)

R^2 tells us how much better we can do in predicting observation by using the model and computing the simulation by just using the mean observation as a predictor (Jamieson et al. 1998).

$$R^2 = 1 - \frac{\sum\limits_{n}^{i=1}(O_i - S_i)^2}{\sum\limits_{n}^{i=1}(O_i = \bar{O})^2} \tag{2.2}$$

R^2 ranges from 0 to 1, with higher values indicating less error variance, and typically values greater than 0.5 are considered acceptable (Santhi et al. 2001; Van Liew et al. 2003; Moriasi et al. 2007).

Root Mean Square Error per Observation (RMSE/obs)

It gives the general standard deviation of the model prediction error per observation (Jamieson et al. 1998).

Table 2.6 Goodness of fit test between measured and projected meteorological data by four models in El-Monofia governorate

	BCC-CSM1-1 model			CCSM4 model		
	d-stat	R^2	RMSE/obs	d-stat	R^2	RMSE/obs
RCP2.6	0.827	0.624	0.246	0.882	0.677	0.212
RCP4.5	0.830	0.639	0.240	0.877	0.659	0.213
RCP6.0	0.832	0.641	0.241	**0.883**	**0.675**	**0.209**
RCP8.5	0.836	0.646	0.239	0.878	0.665	0.213
	GFDL model			MIROC5 model		
	d-stat	R^2	RMSE/obs	d-stat	R^2	RMSE/obs
RCP2.6	0.828	0.626	0.249	0.841	0.620	0.227
RCP4.5	0.830	0.632	0.249	0.845	0.630	0.225
RCP6.0	0.833	0.638	0.251	0.846	0.632	0.225
RCP8.5	0.832	0.640	0.248	0.848	0.639	0.224

$$\text{RMSE} / \text{obs} = \sqrt{\left(\frac{\sum\limits_{n}^{i=1}(S_i - O_i)^2}{n}\right)} \tag{2.3}$$

Where n represents the number of observed and simulated values used in comparison.

Regarding to El-Monofia governorate, the averages goodness of fit test for each RCP scenario for the selected four GCMs models are shown in Table 2.6. The results revealed that RCP6.0 scenario from CCSM4 model was suitable and more accurate to project the meteorological variables because it achieved the highest d-stat, R^2 and the lowest RMSE/obs.

Similar results were found for El-Fayoum governorate, where RCP6.0 scenario from CCSM4 model has the highest d-stat and R^2 values and lowest RMSE/obs value (Table 2.7).

Regarding to Qena governorate (Table 2.8), the goodness of fit test revealed that RCP6.0 climate change scenario developed by CCSM4 model produced the highest d-stat and R^2 values and the lowest RMSE/obs values.

The results existed in Tables 2.6, 2.7, and 2.8) clearly showed that the RCP6.0 scenario from CCSM4 model is more suitable for the studied governorates to be used in projecting weather data in 2030. The CCSM4 model has the highest horizontal resolution, namely $1.25° \times 0.94°$. Furthermore, RCP6.0 is a stabilization scenario, where a range of technologies and strategies for reducing greenhouse gas emissions will be applied (Fujino et al. 2006; Hijioka et al. 2008). Similar results were obtained for several governorates in Egypt (Morsy 2015 and Morsy et al. 2015; Sayad et al. 2015).

BISm model (Snyder et al. 2004) was used to calculate ETo (mm/day) values in the studied governorates in 2030 using projected data by the RCP6.0 scenario from

Table 2.7 Goodness of fit between measured and projected weather data by four models in El-Fayoum governorate

	BCC-CSM1-1 model			CCSM4 model		
	d-stat	R^2	RMSE/obs	d-stat	R^2	RMSE/obs
RCP2.6	0.867	0.614	0.223	0.892	0.686	0.200
RCP4.5	0.872	0.625	0.215	0.889	0.677	0.200
RCP6.0	0.874	0.629	0.215	**0.891**	**0.690**	**0.198**
RCP8.5	0.875	0.636	0.216	0.891	0.688	0.198
	GFDL model			MIROC5 model		
	d-stat	R^2	RMSE/obs	d-stat	R^2	RMSE/obs
RCP2.6	0.865	0.628	0.224	0.885	0.658	0.203
RCP4.5	0.866	0.632	0.224	0.887	0.663	0.201
RCP6.0	0.869	0.646	0.225	0.890	0.669	0.199
RCP8.5	0.869	0.646	0.222	0.892	0.677	0.197

Table 2.8 Averages goodness of fit test between measured and projected weather data by four models in Qena governorate

	BCC-CSM1-1 model			CCSM4 model		
	d-stat	R^2	RMSE/obs	d-stat	R^2	RMSE/obs
RCP2.6	0.844	0.576	0.224	0.855	0.621	0.209
RCP4.5	0.851	0.588	0.215	0.856	0.620	0.206
RCP6.0	0.852	0.588	0.216	**0.860**	**0.630**	**0.202**
RCP8.5	0.856	0.601	0.216	0.856	0.619	0.207
	GFDL model			MIROC5 model		
	d-stat	R^2	RMSE/obs	d-stat	R^2	RMSE/obs
RCP2.6	0.835	0.582	0.228	0.850	0.595	0.225
RCP4.5	0.837	0.594	0.225	0.857	0.610	0.218
RCP6.0	0.842	0.600	0.229	0.845	0.584	0.229
RCP8.5	0.847	0.618	0.220	0.857	0.618	0.219

CCSM4 model. The results in Tables 2.9, 2.10, and 2.11) present the projected monthly weather elements in El-Monofia, El-Fayoum and Qena governorates in 2030, namely solar radiation (SR, MJ/m^2/day), maximum temperature (TMax, °C), minimum temperature (Tmin, °C), wind speed (WS, m/s) and evapotranspiration (ETo, mm/day) values.

Water Requirements for the Studied Crops in 2030

The BISm model (Snyder et al. 2004) was used to calculate water requirements for wheat, faba bean in 2029/2030 season and for maize in 2030 season. Water requirements values for wheat irrigated with surface irrigation (60 % application efficiency), under raised beds cultivation (20 % lower than applied amount under

Table 2.9 Monthly mean values for climate elements projected by RCP6.0 scenario from CCSM4 model and ETo values in 2030 in El-Monofia governorate

	SR (MJ/m²/day)	Tmax(°C)	Tmin (°C)	WS (m/s)	ETo (mm/day)
January	13.4	20.5	10.4	2.2	2.6
February	18.1	20.6	9.9	3.2	3.9
March	22.6	28.7	12.4	3.2	6.3
April	25.9	30.1	15.3	3.2	6.9
May	28.0	36.6	20.3	2.7	8.3
June	30.5	39.0	25.0	3.0	9.4
July	29.2	39.8	25.7	2.9	9.0
August	27.6	39.8	26.4	3.1	9.0
September	24.2	35.8	22.8	3.0	7.4
October	19.4	31.4	19.6	2.6	5.3
November	14.0	28.4	17.9	2.4	4.0
December	12.9	22.3	11.8	2.1	2.9

Table 2.10 Monthly mean values for climate elements projected by RCP6.0 scenario from CCSM4 model and ETo in 2030 in El-Fayoum governorate

	SR (MJ/m²/day)	Tmax(°C)	Tmin (°C)	WS(m/s)	ETo (mm/day)
January	14.7	19.7	8.6	3.1	3.1
February	19.2	22.5	7.9	3.2	4.1
March	23.7	28.9	11.4	3.3	5.9
April	26.5	29.4	13.9	3.3	6.4
May	27.9	36.9	19.2	3.0	7.9
June	30.5	38.4	24.0	3.4	8.8
July	29.3	38.9	24.1	3.4	8.7
August	27.9	38.8	24.9	3.7	8.7
September	24.9	34.8	20.7	3.5	7.2
October	19.8	29.9	18.0	3.0	5.2
November	16.0	27.0	12.9	2.7	4.1
December	13.8	21.1	8.9	2.4	2.9

surface irrigation) and irrigated with sprinkler system (80 % application efficiency) are presented in Table 2.12.

Water requirements for faba bean (Table 2.13) and maize (Table 2.14) irrigated with surface irrigation, under raised beds cultivation and drip system (90 % application efficiency) were also calculated.

Conclusion

There is a strong relation between weather in a region and water requirements of the cultivated crops in it. Thus, it is very important to define annual and seasonal climate of Egypt with respect to maximum and minimum temperatures, wind speed

Table 2.11 Monthly mean values for climate elements projected by RCP6.0 scenario from CCSM4 model and ETo in 2030 in Qena governorate

	SR (MJ/m²/day)	Tmax(°C)	Tmin (°C)	WS (m/s)	ETo (mm/day)
January	15.9	23.7	11.1	2.8	4.5
February	20.3	26.8	9.8	3.4	6.1
March	24.5	32.2	14.0	3.2	7.8
April	26.3	33.1	16.9	3.3	8.7
May	28.8	39.6	22.6	3.7	11.0
June	30.3	41.4	26.6	4.2	12.5
July	29.1	41.8	27.2	4.5	12.9
August	28.1	41.5	28.2	4.3	12.3
September	25.7	39.3	23.3	4.8	11.6
October	21.9	33.8	20.8	3.5	8.2
November	16.8	30.5	17.5	3.3	6.2
December	14.8	24.0	11.0	2.8	4.3

Table 2.12 Water requirements (WR, m³/ha) for wheat under different irrigation systems in 2029/2030

	WR under surface irrigation	WR under raised beds	WR under sprinkler system
Lower Egypt	7721	6177	5791
Middle Egypt	8823	7058	6617
Upper Egypt	11,162	8929	8371
Average	9235	7388	6926

Table 2.13 Water requirements (WR, m³/ha) for faba bean under different irrigation systems in 2029/2030

	WR under surface irrigation	WR under raised beds	WR under drip system
Lower Egypt	6550	5240	4367
Middle Egypt	7583	6067	5056
Upper Egypt	9767	7813	6511
Average	7967	6373	5311

Table 2.14 Water requirements (WR, m³/ha) for maize under different irrigation systems in 2030

	WR under surface irrigation	WR under raised beds	WR under drip system
Lower Egypt	11,700	9360	7800
Middle Egypt	13,317	10,653	8878
Upper Egypt	16,200	12,960	10,800
Average	13,739	10,991	9159

and solar radiation. These are the key elements for calculating water requirements for cultivated crops. Our analysis indicated that the gradient of maximum temperature intensifies during summer and spring seasons over northern part of Egypt and decline over the rest of Egypt. On the contrary, the gradient of minimum temperature intensifies during winter and autumn seasons over northern part of Egypt and decline over the rest of Egypt. The analysis also indicated that there is a strong gradient of wind over northern part according to Mediterranean depression in winter season. Furthermore, downward short wave radiation decreases gradually from south to north according to the apparent position of sun and reaches its maximum value in the summer season.

To achieve higher level of certainty in the calculation of water requirements for the studied crops in 2030, we implemented statistical analysis by comparing between RCPs scenarios developed from four global models (BCC-CSM1-1, CCSM4, GFDL-ESM2G and MIROC5) and measured weather data for a 30-year period, where goodness of fit test was applied. Our analysis indicated that RCP6.0 scenario from CCSM4 model has the highest agreement between measured and projected weather data. Thus, the data from this scenario was used to develop more accurate values of water requirements for the studied crops grown in the three main regions in Egypt in 2029/2030 season. The implemented methodology in this chapter could be a base for projection of climate change effects on different economic sectors.

References

Abtew, W., & Melesse, A. (2013). Climate change and evapotranspiration In *Evaporation and evapotranspiration: Measurements and estimations*. Dordrecht: Springer Science Business Media Dordrecht. doi: 10.1007/978-94-007-4737-113

Allen, R. G., Jensen, M. E., Wright, J, L., & Burman, R. D. (1989). Operational estimate of reference evapotranspiration. *Agronomy Journal, 81*, 650–662.

Allen, R. G., Pereira, L. S., Raes, D., & Smith, M. (1998). Crop evapotranspiration: Guideline for computing crop water requirements. *FAO, 56*, D05109.

Clarke, L. E., Edmonds, J. A., Jacoby, H. D., Pitcher, H., Reilly, J. M., & Richels, R. (2007). *Subreport 2.1a of synthesis and assessment product 2.1*. Washington, DC: Climate Change Science Program and the Subcommittee on Global Change Research.

EEAA. (2010). *Egypt second national communication under the United Nations framework convention on climate change*. Cairo: Egypt Environmental Affairs Agency.

El-Fandy, M. G. (1948). Baroclinic low of Sybrus. *Quarterly Journal of the Royal Meteorological Society, 72*, 291–306.

El-Massah, S., & Omran, G. (2014). *Would climate change affect the imports of cereals? The case of Egypt", handbook of climate change adaptation*. Berlin: Springer.

Fujino, J., Nair, R., Kainuma, M., Masui, T., & Matsuoka, Y. (2006). Multi-gas mitigation analysis on stabilization scenarios using AIM global model. Multi-gas mitigation and climate policy. *The Energy Journal Special, 27*, 343–353.

Hasanean, H. M., & Abdel Basset, H. (2006). Variability of summer temperature over Egypt. *International Journal of Climatology, 26*, 1619–1634.

Hijioka, Y., Matsuoka, Y., Nishimoto, H., Masui, M., & Kainuma, M. (2008). Global GHG emissions scenarios under GHG concentration stabilization targets. *Journal of Global Environmental Engineering, 13*, 97–108.

IPCC. (2013). Summary for policymakers. In T. F. Stocker, G. K. Qin, M. Plattner, S. Tignor, J. Boschung, A. Nauels, Y. Xia, V. Bex, & P. M. Midgley (Eds.), *Climate change. The physical science basis* (Contribution of working group I to the fifth assessment report of the intergovernmental panel on climate change). Cambridge/New York: Cambridge University Press.

Jamieson, P. D., Porter, J. R., Goudriaan, J., Ritchie, J. T., Keulen, H., & Stol, W. (1998). A comparison of the models AFRCWHEAT2, CERES-Wheat, Sirius, SUCROS2 and SWHEAT with measurements from wheat grown under drought. *Field Crops Research, 55*, 23–44.

Khalil, A. A. (2013). Effect of climate change on evapotranspiration in Egypt. *Researcher, 5*(1), 7–12.

Monteith, J. L. (1965). Evaporation and environment. In G. E. Fogg (Ed.), *Symposium of the society for experimental biology: The state and movement of water in living organisms* (Vol. 19, pp. 205–234). New York: Academic Press, Inc.

Moriasi, D. N., Arnold, J. G., Van Liew, M. W., Bingner, R. L., Harmel, R. D., & Veith, T. L. (2007). Model evaluation guidelines for systematic quantification of accuracy in watershed simulations. *American Society of Agricultural and Biological Engineers, 50*(3), 885–900.

Morsy, M. (2015). *Use of regional climate and crop simulation models to predict wheat and maize productivity and their adaptation under climate change*. PhD thesis, Faculty of Science, Al-Azhar University.

Morsy, M., Sayad, T. A., & Ouda, S. (2015). Potential evapotranspiration under present and future climate. In *Management of climate induced drought and water scarcity in Egypt: Unconventional solutions*. Cham: Springer Publishing House.

Ouda, S., Abd El-Latif, K., & Khalil, F. (2016). Water requirements for major crops. In *Major crops and water scarcity in Egypt*. Cham: Springer Publishing House.

Rao, K. P. C., Ndegwa, W. G., Kizito, K., & Oyoo, A. (2011). Climate variability and change: Farmer perceptions and understanding of intra-seasonal variability in rainfall and associated risk in semi-arid Kenya. *Experimental Agriculture, 47*, 267–291.

Riahi, K., Grübler, A., & Nakicenovic, N. (2007). Scenarios of long-term socioeconomic and environmental development under climate stabilization Greenhouse Gases-Integrated Assessment. *Special Issue of Technological Forecasting and Social Change, 74*(7), 887–935.

Santhi, C., Arnold, J. G., Williams, J. R., Dugas, W. A., Srinivasan, R., & Hauck, L. M. (2001). Validation of the SWAT model on a large river basin with point and nonpoint sources. *Journal of American Water Resources Association, 37*(5), 1169–1188.

Sayad, T. A., Ouda, S., Morsy, M., & El Hussieny, F. (2015). Robust statistical procedure to determine suitable scenario of some CMIP5 models for four locations in Egypt. *Global Journal of Advanced Research, 2*(6), 1009–1019.

Shahid, S. (2011). Impact of climate change on irrigation water demand of dry season Boro Rice in Northwest Bangladesh. *Climatic Change, 105*, 433–453. http://dx.doi.org/10.1007/s10584-010-9895-5

Smith, S. J., & Wigley, T. M. L. (2006). Multi-gas Forcing stabilisation with the MiniCAM. *The Energy Journal Special, 3*, 373–392.

Snyder, R. L., Orang, M., Bali, K., & Eching, S. (2004). *Basic irrigation scheduling (BIS)*. http://www.waterplan.water.ca.gov/landwateruse/wateruse/Ag/CUP/Californi/Climate_Data_010804.xls

Solomon, S., Qin, D., Manning, M., Chen, Z., Marquis, M., Averyt, K., Tignor, M., & Miller, H. (Eds.). (2007). *Climate change 2007: The physical science basis* (Contribution of working group I to the fourth assessment report of the intergovernmental panel on climate change, p. 996). Cambridge/New York: Cambridge University Press.

Van Liew, M. W., Arnold, J. G., & Garbrecht, J. D. (2003). Hydrologic simulation on agricultural watersheds: Choosing between two models. *Transactions of the American Society of Agricultural Engineers, 46*(6), 1539–1551.

Van Vuuren, D. P., Eickhout, B., Lucas, P. L., & den Elzen, M. G. J. (2006). Long-term multi-gas scenarios to stabiliseradiative forcing – Exploring costs and benefits within an integrated assessment framework. Multi-gas mitigation and climate policy. *The Energy Journal Special, 27*, 201–233.

Van Vuuren, D. P., den Elzen, M. G. J., Lucas, P. L., Eickhout, B., Strengers, B. J., Van Ruijven, B., Wonink, S., & Van Houdt, R. (2007). Stabilizing greenhouse gas concentrations at low levels: An assessment of reduction strategies and costs. *Climatic Change, 81*, 119–159.

Wayne, G. P. (2013). *The Beginner's guide to representative concentration pathways*. Skeptical Science, Version 1.0. http://www.skepticalscience.com/d°Cs/RCP_ Guide.pdf

Willmott, C. J. (1981). On the validation of models. *Physical Geography, 2*, 184–194.

Wise, M. A., Calvin, K. V., Thomson, A. M., Clarke, L. E., Bond-Lamberty, B., Sands, R. D., Smith, S. J., Janetos, A. C., & Edmonds, J. A. (2009). Implications of limiting CO2 concentrations for land use and energy. *Science, 324*, 1183–1186.

Zohry, A. A., & Ouda, S. (2016a). Crops intensification to face climate induced water scarcity in Nile delta region. In *Management of climate induced drought and water scarcity in Egypt: Unconventional solutions*. Cham: Springer Publishing House.

Zohry, A. A., & Ouda, S. (2016b). Upper Egypt: Management of high water consumption crops by intensification. In *Management of climate induced drought and water scarcity in Egypt: Unconventional solutions*. Cham: Springer Publishing House.

Chapter 3
Projections of the Egyptian Population

Huda Alkitkat

Introduction

In accordance with the United Nations' agenda for the 2030 for sustainable development, namely the second goal: "End hunger, achieve food security and improved nutrition and promote sustainable agriculture" (United Nations 2015). The relationship between food security and population growth is a very critical and hot issue that has been raised in most of the developing countries with overpopulation problem like Egypt. Population represents the demand side for agricultural commodities. Thus, population and population characteristics, including growth, age, and sex is the main influencing factors in the consumption of agricultural commodities (USAID 2006). Egypt's population has been growing rapidly, however it is concentrated in a narrow strip along the Nile River. As population grows, the amount of land needed for housing and businesses rises, and the amount of land for agriculture falls. So Egypt can produce less of its own food, as time goes on (CIA 2016).

This chapter presented Egypt's demographic profile, in addition to projection for Egypt population in 2030. The demographic profile of Egypt is affecting the consumption side of production-consumption food gaps. Furthermore, assessment of future food gap is connected with continues population increase with high growth rates.

H. Alkitkat (✉)
Fellow of Academy of Scientific Research and Technology, Cairo, Egypt
e-mail: hudaalkitkat.phd@gmail.com

© Springer International Publishing AG 2017
S. Ouda et al., *Future of Food Gaps in Egypt*, SpringerBriefs in Agriculture,
DOI 10.1007/978-3-319-46942-3_3

Egypt Demographic Profile

Population Size

The population of Egypt accounts for more than 20 % of the population in the Arab world (PRB 2015). The Egyptian population size increased almost by more than 1.5 times during the last three decades, between 1986 and 2013, from about 48 million to about 85 million in 2013) (CAPMAS 2014). Egypt consists of 27 governorates (Fig. 3.1) and for the purpose of this study the population of Egypt were grouped into three main regions: Lower Egypt region that includes: Alexandria, Behera, Gharbia, Kafr-El Sheikh, Dakahlia, Damietta, Sharkia, Ismailia, Port Said, Suez, Menoufia, Kalubia, Cairo, North Sinai and South Sinai. Middle Egypt region included Giza, Beni Swief, Fayoum, Menia, and Marsa Matrouh. Upper Egypt region included Assuit, Suhag, Qena, Luxor, Aswan, Red Sea, Elwadi Elgadeed.

In 1986, Lower Egypt region had the highest percentage of total population of Egypt as its population size was about 31.0 million, Cairo and Dakahlia were the most populated governorates in that region, the gap between Lower Egypt region and the Upper and Middle Egypt regions was wide as the population size of Middle region was about 9.5 million and the population size of Upper Egypt region was only about 8.0 million inhabitants in that year (CAPMAS 1987).

In 2013 (CAPMAS 2014), the Lower Egypt region also was the most populated region in comparison to the Middle and Upper Egypt regions, where its population size increased by about 68 % along the previous three decades from 1986 to 2013 to reach about 52 million. Cairo and Sharkia governorates were the most populated governorates with about 9.0 million and 6.0 million inhabitants, respectively. The population of Middle and Upper Egypt regions increased with higher percentages, in comparison to the Lower Egypt region during the same period with about 94 % and 82 %, respectively to reach about 18.4 and 14.4 million inhabitants (Table 3.1).

Birth and Death Rates

Demographers believe that societies pass through cycles divided into four stages (Demographic transition); each stage has its characteristics. The first stage: low rate of population growth, high birth rate and high death rate. Second stage: relative stability of the birth rate at its high level during the first stage and gradual decline of the death rate. The third stage: decline in the birth rate and gradually the death rate remarkably declines to its lowest level. The fourth stage: stable and balanced population growth, since successive population plans and programs succeed in achieving their objectives and reach their goals of reducing the birth rate to its lowest possible level. Consequently, the difference between the birth and death rates becomes negligible and does not result in a significant population growth (Nassar 2006).

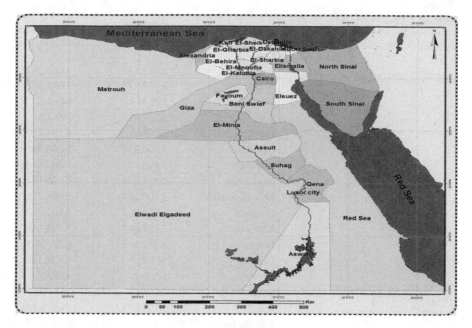

Fig. 3.1 Map of Egypt governorates

Currently, Egypt is in the third stage, the crude birth rate[1] declined during the past three decades from about 40.0 per thousand to 31.0 per thousand, the governorates of Lower Egypt region had crude birth rates between 29.4 per thousand and 49.2 per thousand in 1986 and this bounds declined to 20.5 and 35.0 per thousand in 2013. North Sinai had the highest crude birth rate, i.e. 49.2 per thousand, in 1986 and 35.0 per thousand in 2013 (Table 3.2) (CAPMAS 1986, 2013).

The governorates of the Middle Egypt region had crude birth rates between about 40.0 per thousand and 60.0 per thousand in 1986. Marsa Matrouh governorate had the highest crude birth rate of 60.0 per thousand. In 2013 crude birth rate's bounds of the Middle region declined to be 31.5 and 50.9 per thousand, and Marsa Matrouh also had the highest rate (Table 3.2). The governorates of the Upper Egypt region had crude birth rates between about 37.4 per thousand and 51.0 per thousand in 1986; Qena had the highest crude birth rate of 5.1 per thousand. In 2013, crude birth rate bounds of Upper Egypt region declined to be 28.7 and 33.9 per thousand, and Assuit had the highest rate (Table 3.2).

Crude death rate[2] is also declined in Egypt during the last three decades from 9.6 per thousand in 1986 to 6.0 per thousand in 2013. The governorates of Lower Egypt region had crude death rates between 4.3 per thousand and 10.2 per thousand in

---

[1] Crude birth rate is defined as the number of live births occurring among the population of a given geographical area during a given year, per 1000 mid-year total population of the given geographical area during the same year.

[2] Crude death rate is defined as the number of deaths occurring among the population of a given geographical area during a given year, per 1000 mid-year total population of the given geographical area during the same year.

Table 3.1 Population
distribution by place of
residence in 1986 and 2013 in
Egypt

Place of residence	Population 1986[a]	2013[b]
Lower Egypt		
Cairo	6,052,836	9,002,783
Alexandria	2,917,327	4,658,381
Port-Said	399,793	646,461
Suez	326,820	599,320
Damietta	741,264	1,284,710
Dakahlia	3,500,470	5,748,965
Sharkia	3,420,119	6,242,810
Kalyoubia	2,514,244	4,926,148
Kafr-El-Sheikh	1,800,129	3,054,770
Gharbia	2,870,960	4,592,222
Menoufia	2,227,087	3,799,149
Behera	3,257,168	5,563,465
Ismailia	544,427	1,128,373
North Sinai	171,505	415,532
South Sinai	28,988	163,092
Total	30,773,137	51,826,181
Middle Egypt		
Giza	3,700,054	7,291,017
Beni-Swief	1,442,981	2,727,614
Fayoum	1,544,047	3,021,448
Menia	2,648,043	4,930,641
Matrouh	160,567	417,294
Total	9,495,692	18,388,014
Upper Egypt		
Assuit	2,223,034	4,062,821
Suhag	2,455,134	4,404,545
Qena	2,252,315	2,918,086
Aswan	801,408	1,374,985
Luxor	–	1,104,858
Red Sea	90,491	332,741
Elwadi Elgadeed	113,838	216,751
Total	7,936,220	14,414,787
Total Egypt	48,205,049	84,628,982

Source:
[a]Central Agency of Public Mobilization and Statistics,
annual statistical year book 1987
[b]Central Agency of Public Mobilization and Statistics,
annual statistical year book 2014

1986 and this bound declined to 3.9 and 8.8 per thousand in 2013. Monofia had the
highest crude death rate, i.e. 10.2 per thousand in 1986 and Cairo had the highest
rate, i.e. 8.8 per thousand in 2013 (Table 3.2). The governorates of the Middle Egypt

Table 3.2 Crude birth rate and crude death rate by place of residence in 1986 and 2103

Place of residence	Crude birth rate		Crude death rate	
	1986[a]	2013[b]	1986[a]	2013[b]
Lower Egypt				
Cairo	30.9	28.5	9.6	8.8
Alexandria	31.4	28.2	7.9	7.9
Port-Said	29.4	24.4	6.0	6.4
Suez	37.3	26.9	6.9	5.8
Damietta	37.1	29.4	7.1	6.4
Dakahlia	37.8	28.3	7.2	6.4
Sharkia	41.0	30.1	9.0	5.6
Kalyoubia	39.0	28.5	8.2	5.0
Kafr-ElSheikh	41.2	30.5	8.2	5.4
Gharbia	36.4	28.4	8.1	6.5
Menoufia	38.7	30.7	10.2	5.9
Behera	40.0	32.5	9.0	5.2
Ismailia	39.6	34.0	7.5	5.8
North Sinai	49.2	35.0	6.7	5.0
South Sinai	45.2	20.5	4.3	3.9
Middle Egypt				
Giza	40.1	31.5	10.1	5.6
Beni-Suef	48.6	35.5	11.0	5.4
Fayoum	50.8	35.7	10.5	4.7
Menia	46.2	33.6	12.0	5.2
Matrouh	60.1	50.9	8.0	5.0
Upper Egypt				
Assuit	45.5	33.9	10.8	6.0
Suhag	47.3	33.3	13.4	5.4
Qena	51.0	32.5	10.8	5.3
Aswan	45.0	31.2	13.3	5.4
Luxor	–	29.6	–	5.5
Red Sea	41.2	29.4	5.1	4.2
ElWadi ElGidid	37.4	28.7	6.1	4.0
Egypt average	40.1	31.0	9.6	6.0

Source:
[a]Central Agency of Public Mobilization and Statistics, annual birth and death statistical bulletin 1986
[b]Central Agency of Public Mobilization and Statistics, annual birth and death statistical bulletin 2013

region had death rates between about 8.0 per thousand and 12.0 per thousand in 1986. Menia governorate had the highest crude death rate of 12.0 per thousand. In 2013, crude death rate's bound of Middle Egypt region declined to be 4.7 and 5.6 per thousand and Giza had the highest rate (Table 3.2). The governorates of Upper Egypt region had crude death rates between about 5.1 per thousand and 13.4 per thousand in 1986. Suhag governorate had the highest crude death rate of 13.4 per thousand. In 2013 crude death rate's bounds of the Upper region declined to be 4.0 and 6.0 per thousand, and Assuit had the highest rate (Table 3.2).

Egypt Population Projection in 2030

Population projections may be defined as the numerical outcome of a particular set of assumptions regarding the future population. It is the basic tools for a wide range of decision makers and planners in many sectors: education, health, manpower, food supply and many other services in any country. One of the main roles that population projections play is to support planners and decision makers to measure and fit the future food gap. This section provides population projection for Egypt in 2030, which will be used to calculate national production-consumption for the selected crops.

Model of population projection is produced to project Egypt population by age and place of residence to the year of 2030 using Spectrum 5, 2016 program, where cohort-component method was used. Cohort–component method was first employed in producing global population projections by Notestein (1945). According to this method, the population is divided into age-sex groups (birth cohort) and accounts separately for fertility, mortality and migration behavior of each cohort along the projection horizon. For the population distribution by age, sex and place of residence at the base year of the projection, CAPMAS estimation for population in mid of the year 2013 was used and the age and sex composition were calculated using the post enumeration survey for the last population census in 2006 (CAPMAS 2008).

For the fertility component, data from series of Egypt demographic and health surveys[3] were used. Furthermore, for the mortality component, data from series of Egypt demographic and health surveys were used to estimate the life expectancy at birth. For the migration component, the results of the last census in Egypt in 2006 were used to calculate the current internal migration between governorates, and then the United Nations' estimates were used to calculate the international migration. The trend scenario has been adopted that assumed fertility and mortality components will continue following the past trend to 2030 and the migration component will be constant as its level in 2006, where the migration is not significant in the population growth in Egypt as much as the fertility component.

The results showed that Egypt population size will increase with about 49 % of its size in 2013 to reach about 126 million in 2030. Lower Egypt region will have the highest percentage of total population of Egypt as its population size will be more than 76.0 million inhabitants, Cairo and Sharkia will be the most populated governorates in that region with population size of more than 11.0 and 9.0 million

---

[3] Egypt demographic and health survey is conducted regularly each 4/5 years on a sample of ever married women age 15–49. The main purpose of the EDHS is to provide detailed information on fertility, family planning, infant and child mortality, maternal and child health and nutrition.

Table 3.3 Population
projection by place of
residence in Egypt in 2030

Place of residence	Population size in 2030
Lower Egypt	
Cairo	11,994,250
Alexandria	6,188,125
Port-Said	860,208
Suez	807,838
Damietta	1,945,973
Dakahlia	8,755,783
Sharkia	9,647,746
Kalyoubia	7,601,266
Kafr-ElSheikh	4,720,344
Gharbia	6,980,000
Menoufia	5,804,627
Behera	8,624,799
Ismailia	1,744,299
North Sinai	650,808
South Sinai	221,629
Total	76,547,695
Middle Egypt	
Giza	10,833,066
Beni-Suef	4,119,260
Fayoum	4,564,334
Menia	7,406,478
Matrouh	665,328
Total	27,588,466
Upper Egypt	
Assuit	6,140,411
Suhag	6,620,888
Qena	4,422,797
Aswan	2,073,674
Luxor	1,660,289
Red Sea	484,671
ElWadi ElGidid	331,846
Total	21,734,576
Egypt Total	125,870,737

Source: Author's estimations

inhabitants respectively. The gap between Lower Egypt region and the Upper and Middle regions will be wide, as the Middle region population size will be about 28.0 million and the Upper region population size will be about 22.0 million inhabitants (Table 3.3).

The results of Egypt population projection showed that the number of working-age population (15–64 years old) will reach more than 75.0 million in 2030, which will represent about 60 % of the total population. Lower Egypt region will have the majority of that age group, i.e. more than 45.0 million. This situation will be good opportunity to the planners and decision makers responsible for solving food gaps problems, where they can prepare well to catch that opportunity by producing some specialized training programs on agricultural techniques, in which those population groups could be involved and be part of the production process of food crops in the future.

Conclusion

Egypt is one of the developing countries that suffer from the overpopulation problem, which result in a large food gaps problem. The population size was about 48 million in 1986 and according to the population projection results this size will increase to reach more than 125 million by 2030. The rapid population growth rate in Egypt will lead to duplication in the population size with about 2.5 times from 1986 to 2030. Population distribution by place of residence showed that Lower Egypt region had the highest population size along the period from 1986 to 2030. Cairo governorate was and will be the most populated governorate in Egypt during the same period. Dakahlia was in the second rank in 1986 and Sharkia will be in the second rank in 2030.

References

Central Agency of Public Mobilization and Statistics (CAPMAS). (1986). *Annual birth and death statistical bulletin 1986.*

Central Agency of Public Mobilization and Statistics (CAPMAS). (1987). *Annual statistical year book 1987.*

Central Agency for Public Mobilization and Statistics (CAPMAS). (2008). *2006 Census for Population and Houses.*

Central Agency of Public Mobilization and Statistics (CAPMAS). (2013). *Annual birth and death statistical bulletin 2013.*

Central Agency of Public Mobilization and Statistics (CAPMAS). (2014). *Annual statistical year book 2014.*

CIA. (2016). *World fact book.* https://www.cia.gov/library/publications/the-world-factbook/geos/print/country/countrypdf_eg.pdf

Nassar, S. (2006). *Policy implication of the demographic dividend (window of opportunity) and its consequences on the labor market: A case study of Egypt" EPDI.*

Notestein, F. W. (1945). Population-The long view. In T. W. Schultz (Ed.), *Food for the world.* Chicago: University of Chicago Press.

Population Reference Bureau (PRB). (2015). *2015 world population data sheet.*

United Nations. (2015). Transforming our world: The 2030 Agenda for Sustainable Development. Resolution adopted by the General Assembly on 25 September 2015.

USAID. (2006). *Impact on Egypt's economy of eliminating domestic support for commodities: Literature review of supply and demand elasticity 2006.*

Chapter 4
Crops Intensification to Reduce Wheat Gap in Egypt

Samiha A.H. Ouda and Abd El-Hafeez Zohry

Introduction

Wheat is grown in Egypt between latitudes 25 °N and 31 °N. Most of the wheat area (57 %) lies in the Nile Delta, but there are small areas in Middle Egypt (18 %) and Upper Egypt (17 %).Wheat is the major staple crop in Egypt and viewed as a strategic commodity, which is considered a main ingredient in the Egyptian diet. Therefore, the consumers have no other choice except consuming the bread since it is still the cheapest food. Consumption of wheat is increasing as a result of the annual population increase approaches 2.0 million/year (Mansour 2012). Egypt continues to have one of the highest wheat per capita consumption levels in the world. In 2011, percentage of total daily calories from wheat was 33 % as indicated by United Nations Food and Agriculture Organization (http://wheatatlas.org/country/EGY). In 1960, Egyptians consumed annually less than 110 kg per capita of wheat. In the 1980s, the wheat supply was enough to provide 175 kg per capita, compared to a world average of less than 60–75 kg per capita (Metz 1990). In 2012, Egyptians, on average, consume up to 200 kg of wheat per capita per year (Aegic 2015).

In wheat growing season of 2012/2013, wheat cultivated area in the old land (clay soil in the Nile Delta and Valley) reached 1,151,121 hectares and the new cultivated area (sandy soil on the borders of the Nile Delta and Valley) was 261,578 hectares with total area equal to 1,412,699 hectares. The total wheat production

S.A.H. Ouda (✉)
Water Requirements and Field Irrigation Research Department, Soils, Water and Environment Research Institute, Agricultural Research Center, Giza, Egypt
e-mail: samihaouda@yahoo.com

A.E.-H. Zohry
Crops Intensifications Research Department, Field Crops Research Institute, Crops Agricultural Research Center, Giza, Egypt
e-mail: abdelhafeezzohry@yahoo.com

© Springer International Publishing AG 2017
S. Ouda et al., *Future of Food Gaps in Egypt*, SpringerBriefs in Agriculture,
DOI 10.1007/978-3-319-46942-3_4

from these areas was 9,153,948 tons as indicated by the Egyptian Ministry of Agriculture and Land Reclamation in 2015 year book. In that year, total population was 84,628,982 inhabitants, which make wheat consumption per inhabitant per year 218.60 kg. Thus, Egypt needed to import an amount of wheat equal to 9,345,948 tons (United State Department of Agriculture, http://www.usda.gov/). This huge amount of importation put a large burden of the government budget. Thus, there is a need to think about solutions to reduce wheat food gap and increase its security.

Two obstacles are expected to confront wheat gap in Egypt: population growth and climate change effect on wheat production and its water requirements. In 2030, Egyptian population is projected to be 125,870,736 inhabitants, which will require producing more wheat grains. Furthermore, previous studies on the effect of climate change on water requirements for wheat in 2030 using AR4 climate change scenarios revealed that its water requirements will increase by 9 % in the Nile Delta and 18 % in both Middle and Upper Egypt (Ouda et al. 2015). Therefore, it is expected that wheat gap will be enlarged in 2030.

Two approaches can be done to increase total wheat production: increase productivity per unit area and increase cultivated area. High yielding wheat cultivars can be used to increase productivity per unit area. However, this approach needs long period of time to breed and test the stability of these new wheat cultivars. Furthermore, increase cultivated area needs to allocate more irrigation water to sandy soils to be cultivated, which cannot be implemented as a result of limitation of our water resources. As we are pressed for time, we need unconventional procedures to save on the applied irrigation water to wheat without any reduction in its productivity. This can be done though using improved agricultural management practices.

The objective of this chapter was to investigate opportunities to increase wheat national production to reduce its current production-consumption gap. The effect of several assumptions was assessed to increase current national wheat production, as well as national wheat production in 2030. It is expected that population will be higher in 2030 and wheat national production will be lower as a result of decease in its cultivated area. Thus, wheat production-consumption gap will be larger.

Current Situation of Wheat Production-Consumption Gap

As we stated before, wheat is cultivated in the clay soil known as old land and on sandy soil known as new land. In clay soil, flat or on narrow rows cultivation using surface irrigation is prevailing. However, it leads to ineffective use of applied nitrogen owing to poor aeration, leaching and volatilization losses. The practice also results in greater crop lodging, lower water use efficiency, and crusting of the soil surface (Majeed et al. 2015).

Table 4.1 showed wheat cultivated area (ha), productivity (ton/ha) and total production (ton) in 2012/2013 growing season in Egypt (collected data from Ministry of Agriculture and Land Reclamation, Egypt).Wheat water requirements (m^3/ha) was estimated by BISm model (Snyder et al. 2004) under surface irrigation (the prevailing irrigation system in the Nile Delta and Valley; old cultivated land) and under sprinkler system (the prevailing system in the new land) with 60 and 80 % application efficiency for old and new lands, respectively. It is shown from Table 4.1 that the total cultivated area in the old land produced 7,962,728 tons and consumed 8,468,698,764 m³.

With respect to wheat cultivated in the sandy soil of Egypt in 2012/2013 growing season, Table 4.2 indicated that the cultivated area is lower than its counterpart in the old land, as well as wheat productivity. Furthermore, the applied water per hectare was lower than its counterpart in the old land as a result of using sprinkler system.

Table 4.3 revealed that total wheat production in the growing season of 2012/2013 was 9,517,752 tons and total water requirements to produce that amount were 9,956,261,583 m³ (Table 4.3).

Table 4.1 Wheat data in the old cultivated area in the growing season of 2012/2013 in Egypt

	Cultivated area (ha)	Productivity (ton/ha)	Production (ton)	Total water requirements (m³)
Lower Egypt	714,623	6.92	4,944,458	5,089,673,083
Middle Egypt	240,305	6.99	1,679,390	1,770,243,764
Upper Egypt	196,193	6.82	1,338,880	1,608,781,917
Total	1,151,121		7,962,728	8,468,698,764

Table 4.2 Wheat data in the sandy soil in the growing season of 2012/2013 in Egypt

	Cultivated area (ha)	Productivity (ton/ha)	Production (ton)	Total water requirements (m³)
Lower Egypt	126,243	6.32	797,462	674,343,194
Middle Egypt	30,545	5.76	176,072	168,761,125
Upper Egypt	104,790	5.55	581,490	644,458,500
Total	261,578		1,555,024	1,487,562,819

Table 4.3 Wheat data (total) in the old cultivated area and sand soil in the growing season of 2012/2013 in Egypt

	Total cultivated area (ha)	Total production (ton)	Total water requirements (m³)
Lower Egypt	840,866	5,741,920	5,764,016,277
Middle Egypt	270,850	1,855,462	1,939,004,889
Upper Egypt	300,983	1,920,370	2,253,240,417
Total	1,412,699	9,517,752	9,956,261,583

Table 4.4 Egyptian population, total wheat consumption and wheat production-consumption gap

	Population in July 2013	Total wheat consumption (ton)	Wheat gap (ton)
Lower Egypt	51,826,181	11,329,203	5,587,284
Middle Egypt	18,388,014	4,019,620	2,164,157
Upper Egypt	14,414,787	3,151,072	1,230,702
Total	84,628,982	18,499,895	8,982,143

Egyptian Population and Wheat Consumption

Table 4.4 revealed that in Lower Egypt, population was estimated to be 51,826,181 inhabitants. This population size needs 11,329,203 tons of wheat to consume and the available amount was only 5,741,920 tons (Table 4.3). Thus, the existed gap in Lower Egypt was 5,587,284 tons (Table 4.4). Accordingly, the gap was 2,164,157 and 1,230,702 tons in Middle and Upper Egypt, respectively, with total amount of 8,982,143 tons.

Suggestions to Increase Wheat Cultivated Area and Production

1. Cultivation on raised beds

Changing cultivation methods from cultivation in basins or on narrow rows (0.6 m in width) to raised beds (1.2 m in width) represents an opportunity here to save on the applied irrigation water to wheat. This method proved to save 20 % of the applied water to wheat under surface irrigation (Abouelenein et al. 2009). Furthermore, this method increased wheat yield by 15 %, as a result of increase in radiation used efficiency, where plants are more exposed to solar radiation; it also increase nitrogen use efficiency and increase water use efficiency (Aboelenein et al. 2011). Majeed et al. (2015)) indicated that raised bed planting of wheat not only saves water but improved fertilizer use efficiency and increase grain yield by 15 % compared to flat planting.

Wheat water requirements (m³/ha) under raised beds cultivation was estimated using BISm model (Snyder et al. 2004) with 20 % saving in the applied water to surface irrigation. The results in Table 4.5 revealed that using raised beds for wheat cultivation can increase wheat productivity per hectare and total wheat production in the old land of Egypt from 7,962,728 tons (Table 4.1) to 9157,138 tons (Table 4.5), with lower water requirements than its counterpart under traditional method.

Under these circumstances, the saved irrigation water can be used to cultivate new area with wheat equal to 306,966 hectares, which can produce 1,863,490 tons of wheat and that can increase the cultivated area of wheat in the new land increase the cultivated area to 1,719,664 hectares (Table 4.6).

Table 4.5 Potential wheat cultivated area, productivity and production under raised beds cultivation in the old land of Egypt

	Cultivated area (ha)	Productivity (ton/ha)	Production (ton)	Total water requirements (m³)
Lower Egypt	714,623	7.96	5,686,127	4,071,738,466
Middle Egypt	240,305	8.04	1,931,299	1,416,195,011
Upper Egypt	196,193	7.85	1,539,713	1,287,025,533
Total	1,151,121		9,157,138	6,774,959,011

Table 4.6 Potential new cultivated area with wheat, its production and total production under raised beds cultivation in Egypt

	Available water to cultivate new land (m³/ha)	New cultivated area (ha)	Production of new area (ton)	Total cultivated area (ha)
Lower Egypt	1,017,934,617	190,566	1,203,784	1,031,432
Middle Egypt	354,048,753	64,081	369,387	334,931
Upper Egypt	321,756,383	52,318	290,318	353,301
Total	1,693,739,753	306,966	1,863,490	1,719,664

Table 4.7 Wheat total production and its food under raised beds cultivation

	Total wheat production (ton)	Wheat food gap (ton)
Lower Egypt	7,687,372	3,641,831
Middle Egypt	2,476,758	1,542,862
Upper Egypt	2,411,521	739,552
Total	12,575,651	5,924,244

Thus, total wheat production can reach 12,575,651 tons (Table 4.7) with the same applied irrigation water presented in Table 4.3. This total amount of wheat production is resulted from wheat production from old cultivated land with 15 % increase in productivity as a result of cultivation on raised beds, production from new area cultivated with wheat as a result of irrigation water availability and production from sandy soil (Table 4.7). Thus, changing wheat cultivation to raised beds can reduce wheat gap to 5,924,244 tons (Table 4.7).

2. Irrigation of wheat in the old land with sprinkler system

Sprinkler irrigation is an advanced irrigation technique for water-saving and fertigation and accurately controlling irrigation time and water amounts (Li and Rao 2003). Sprinkler system is one of the useful technologies to increase crop production and water productivity (Liua and Kanga 2006). Changing irrigation system from surface to sprinkler, where application efficiency increases from 60 to 80 % could also enable farmers to save 20 % of irrigation water for wheat (Taha 2012). As a result, wheat productivity in tons will increase by 18 % (Ibrahim et al. 2012).

Water requirements for wheat (m³/ha) was estimated using BISm model (Snyder et al. 2004) under sprinkler system with 80 % application efficiency and presented in Table 4.8. The results in that table indicated that using sprinkler system to irrigate old cultivated area of wheat can increase its total production to 9,396,020 tons as a result of increase in wheat productivity by 18 %. Furthermore, availability of irrigation water after changing the irrigation system can add 383,707 hectares to be cultivated with wheat, which will make the total cultivated area is 1,796,405 hectares (old, new and added lands) (Table 4.8).

Using sprinkler system to irrigate all the cultivated area of wheat can result in increase wheat total production to b 13,280,406 tons and that can reduce wheat production-consumption gap to be 5,219,490 tons (Table 4.9).

3. Wheat intercropping with other crops

Another useful technique to increase productivity per unit area is intercropping, where one crop share its life cycle or part of it with another crop (Eskandari et al. 2009). This practice can be used as a way to improve soil fertility, increase land

Table 4.8 Potential wheat production under sprinkler irrigation in the old and new lands of Egypt

	Old land production (ton)	Added cultivated area (ha)	Production of added area (ton)	Total cultivated area (ha)
Lower Egypt	5,834,460	238,208	1,504,730	1,079,074
Middle Egypt	1,981,680	80,102	461,734	350,951
Upper Egypt	1,579,879	65,398	362,898	366,381
Total	9,396,020	383,707	2,329,362	1,796,406

Table 4.9 Wheat total production and its gap under sprinkler system

	Total production (ton)	Wheat food gap (ton)
Lower Egypt	8,136,652	3,192,551
Middle Egypt	2,619,487	1,400,133
Upper Egypt	2,524,267	626,806
Total	13,280,406	5,219,490

productivity and save on the applied irrigation water (Kamel et al. 2010), as well as increase water productivity as a result of using less water to irrigate two crops (Andersen 2005). The conventional ways of intensifying crop production are vertical and horizontal expansions. Intercropping offers two additional dimensions, time and space. The intensification of land and resources use in space dimension is an important aspect of intercropping. It enhanced the efficient use of light as two or more species that occupy the same land during a significant part of the growing season and have different pattern of foliage display (Francis 1986). Furthermore, different rooting patterns can explore a greater total soil volume because of the roots being at different depths (Francis 1986). These differences in foliage display and rooting patterns create the space dimension of intercropping (Dunn et al. 1999).

Many successful intercropping systems with wheat were proved to increase wheat land and water productivities in Egypt, such as wheat intercropping with tomato or sugar beet. Relay intercropping cotton with wheat was also implemented. These systems should be cultivated on raised beds. Furthermore, wheat intercropped with sugarcane and wheat interplanted under young fruit trees proved to increase wheat cultivated area without any reduction in the productivity of the main crop. Thus, using these intercropping systems can increase wheat national cultivated area with a percentage of the cultivated area of under these crops.

Intercropping wheat with tomato (Fig. 4.1) is implemented by cultivating tomato in the end of September on raised beds with 100 % planting density. After 45 days, wheat seeds are sown in four rows, where tomato plants will be locate between two wheat rows from each side on the top of the raised beds, with 75 % planting density, compared to wheat sole planting. Under this system, wheat plants use the applied amounts of water and fertilizer to tomato and tomato continued to give fruits until the end of March. On the contrary, when tomato is planted solely it finish its life cycle in the end of December. The main benefit of this system occurs for tomato,

Fig. 4.1 wheat intercropped with tomato

Fig. 4.2 Cotton relay intercropping with wheat

where wheat plants protect tomato plants from low temperature in January and February (Abd El-Zaher et al. 2013). Exposing tomato plants to low temperature reduce pollen production, shed, viability and tube growth (Fernandez-Munoz et al. 1995). Pressman et al. (1997) indicated that higher water use efficiency is obtained as a result of implementing this system. Tomato grows tap deep strong root systems, which facilitate the absorption of soil moisture deeper than wheat root system. Furthermore, the roots of tomato leave the soil in a good mechanical condition. In addition, the roots of tomato plants contain more fertilizing elements than those of most other crops (Pressman et al. 1997). For that reason, wheat productivity increased under this system to be 80 % of its sole planting, although 75 % of its planting density was cultivated.

Another successful intercropping system was reported: relay intercropping cotton on wheat (Fig. 4.2), where wheat is cultivated in November on the top of raised beds and in following March cotton is cultivated in on both edges of the raised beds. Thus, cotton shares 2 month of its life cycle with wheat before wheat harvest occur in April. The benefit of this system is to increase wheat cultivated area by the area assigned to be cultivated by cotton. Furthermore, planting density for wheat under this system in 80 % of its recommended density and it attain 90 % of its yield under sole planting (Zohry 2005).

Relay intercropping cotton on wheat system (Fig. 4.2) is characterized by three main phases: (1) wheat vegetative stage grown from November till March; (2) intercropping of wheat (reproductive stage) and cotton (seedling stage) from March till April, and (3) sole cotton (vegetative and reproductive stage) from April till September (Zohry 2005). The two component crops in the system interact directly only during the second phase; however the physiology, ecology and productivity of the relay intercropping system are determined by the spatial architecture and temporal dynamics of the leaf canopy and the root systems during the whole growing cycle (Zhang et al. 2007). Furthermore, Zhang et al. (2008) stated that the N-uptake of cotton was diminished during the intercropping phase, but recovered partially during later growth stages, with low effect on final cotton yield. Furthermore, they also stated that intercrops used more nitrogen per unit production than mono crops, which can reduce environmental risks of leaching to ground water.

Intercropping wheat with sugar beet on raised beds was successfully implemented in Egypt. Sugar beet is an important winter crop in Egypt. Its cultivation resulted in reduction of the assigned area to other winter crops, especially faba been. Thus, to solve the problem of limited fertile land and water resources, several winter crops successfully intercropped with sugar beet, such as wheat and faba bean. In the system of wheat intercropped with sugar beet, sugar beet is cultivated in October and wheat is intercropped on sugar beet 45 days later. Planting density of sugar beet is 100 % and wheat planting density is 50 %, which obtained the same yield of both crops as if they are planted solely. Wheat use the applied irrigation water to sugar beet and both crops are harvest in April (Abou-Elela 2012).The advantage of this system is to increase the cultivated area of wheat by the percentage of assigned area from sugar beet cultivated area (Fig. 4.3).

Another successful intercropping system for wheat is intercropping with fall sugarcane in south Egypt. Nazir et al. (2002) indicated that sugarcane offers a unique potential for intercropping. It is planted in wide rows (100 cm), and takes several months to develop its canopy, during which time the soil and solar energy goes to waste. The growth rate of sugarcane during its early growth stages is slow, with leaf canopy providing sufficient uncovered area for growing of another crop. In this case, wheat will not need any extra irrigation water as it will use the applied

Fig. 4.3 Wheat intercropping with sugar beet

Fig. 4.4 Wheat plants harvested under intercropping with sugarcane

water to sugarcane to fulfill its required needs for water. Furthermore, intercropping wheat on sugarcane provide extra income for farmers during the early growth stage of sugarcane. Under intercropping wheat with sugarcane system (Fig. 4.4), sugarcane is cultivated in September and wheat is cultivated November with 40 % of its recommended planting density then wheat is harvested in April. This system produces 40 % of wheat yield with no reduction in sugarcane yield (Ahmed et al. 2013).

Finally, we suggest intercropping wheat under young evergreen fruit tree (1–3 years old) or deciduous fruit trees. This system can increase the cultivated area with wheat and consequently its national production. Furthermore, this practice is common by Egyptian farmers under rain fed conditions in the winter and under irrigated agriculture for deciduous trees. Intercropping wheat in rows between the fruit trees give an extra economic incentive and also improve land productivity. This practice can be done in young evergreen fruit trees by separation between fruit trees and wheat cultivated area to prevent the runoff of irrigation water to these trees (WOCAT 2016).

Wheat New Added Area and Production Under Intercropping

We assumed that 45 % of the cultivated area with winter tomato will be assigned to be intercropped with wheat and wheat will produce 80 % of its yield under sole planting. Regarding to relay intercropping cotton with wheat, we assumed it will be done on 90 % of the cultivated area of cotton and wheat will produce 90 % of its yield under sole planting. We also assumed that 25 % of sugar beet area will be assigned to be intercropped with wheat and wheat productivity will be 50 % of its yield under sole planting. Regarding to intercropping wheat on sugarcane and under fruit trees, we assumed that 17 % of its cultivated area will be used for intercropping. Furthermore, wheat productivity will be 50 and 70 % under intercropping on sugarcane and under fruit trees, respectively. Table 4.10 indicted that using these wheat five intercropping systems can increase wheat cultivated area by 305,735 hectares.

This potential new cultivated area can produce 1,667,283 ton of wheat, which can contribute in reducing wheat production-consumption gap (Table 4.11).

Thus, if we implemented the suggested wheat intercropping systems and cultivate wheat on raised beds nationally, wheat gap will be reduced from 8,982,143 tons (Table 4.4) to be 4,168,197 tons (Table 4.12).

Similarly, if we assumed that all national wheat cultivated area will be irrigated with sprinkler, food gap will be reduced from 8,982,143 tons (Table 4.4) to 3,463,433 ton (Table 4.13).

Table 4.10 Potential wheat cultivated area under different systems of intercropping

	Potential wheat cultivated area (ha) under intercropping with					
	Tomato	Cotton	Sugar beet	Sugarcane	Fruit trees	Total (ha)
Lower Egypt	16,138	88,607	68,628	–	42,196	215,569
Middle Egypt	17,329	6830	19,395	–	6101	49,654
Upper Egypt	10,463	1290	1902	19,435	7421	40,511
Total	43,930	96,727	89,924	19,435	55,718	305,735

Table 4.11 Potential wheat production under different systems of intercropping

	Wheat potential production (ton) under intercropping with					
	Tomato	Cotton	Sugar beet	Sugarcane	Fruit trees	Total (ton)
Lower Egypt	102,728	634,528	273,028	–	204,367	1,214,652
Middle Egypt	111,417	49,399	77,936	–	29,846	268,598
Upper Egypt	65,690	9114	7463	66,316	35,450	184,034
Total	279,835	693,042	358,427	66,316	269,663	1,667,283

Table 4.12 Wheat total production and its gap under intercropping and raised beds cultivation

	Wheat total production (ton)	Wheat food gap (ton)
Lower Egypt	8,970,685	2,358,518
Middle Egypt	2,793,446	1,226,174
Upper Egypt	2,567,568	583,505
Total	14,331,698	4,168,197

Table 4.13 Wheat total production and the gap under intercropping and sprinkler system		Wheat total production (ton)	Wheat food gap (ton)
	Lower Egypt	9,419,965	1,909,238
	Middle Egypt	2,936,174	1,083,446
	Upper Egypt	2,680,314	470,759
	Total	15,036,453	3,463,443

Effect of Climate Change on Wheat Production

Climate change will alter the normal growing conditions for wheat, which will result in abiotic stress, such as heat and water stresses. Wheat plants are very sensitive to high temperature (Slafer and Satorre 1999), where it is affected by heat stress to varying degrees at different phenological stages, but heat stress during the reproductive phase is more harmful than during the vegetative phase due to the direct effect on grain number and its dry weight (Wollenweber et al. 2003). Pre-anthesis and post-anthesis high temperature may have huge impacts upon wheat growth through reduction in photosynthetic efficiency (Yang et al. 2011).

Exposing wheat plants to high moisture stress depressed seasonal consumptive use and grain yield (Bukhat 2005). During vegetative growth, phyllochron decreases in wheat under water stress (McMaster 1997) and leaves become smaller, which could reduce leaf area index (Gupta et al. 2001) and number of reproductive tillers, in addition to limit their contribution to grain yield (Dencic et al. 2000).

A range of valuable national studies have been carried out and published in Egypt concerning the potential vulnerability of wheat under expected climate change in the future. The productivity of wheat planted in clay soil under sprinkler system in El-Behira governorate, Lower Egypt is expected to be reduce by 21 % as an average over 4 cultivars, compared to wheat yield resulted in 2009/2010 (Abdrabbo et al. 2013). Furthermore, farmer's application of irrigation water in sandy soil using sprinkler system is characterized by large applied irrigation water and application of fertilizer and pesticide by broadcasting on the soil. Therefore, under climate change effects, wheat productivity is expected to be reduced by 30 % for farmer practice (Ouda et al. 2010) and by 38 % under fertigation practice (Ibrahim et al. 2012). In salt affected soil of Demiatte governorate in Lower Egypt and under surface irrigation, wheat yield was reduced by 40 % for farmer's practice (Ouda et al. 2012). Moreover, Khalil et al. (2009) indicated that the productivity of wheat planted in Giza governorate, Middle Egypt in clay soil is expected to decrease by 40 % as an average of 3 cultivars. At the same governorate, where wheat was grown in silty clay soil under sprinkler system, its productivity was reduced by 21 % as an average over 4 cultivars (Abdrabbo 2011). The above results implied that national wheat production will be highly reduced. Moreover, under climate change effect in 2030, Ouda et al. (2015) stated that wheat water requirements will increase by 9 % in the Nile Delta and 18 % in both Middle and Upper Egypt. As a consequence of fixed amount of water allocated to agriculture, it is expected that the national wheat cultivated area will be reduced. In addition, high population increase is expected, which will increase wheat production-consumption gap, as well as food insecurity.

Assessment of Wheat Production-Consumption Gap in 2030

As it was stated in Chap. 2, Egyptian population is projected to be 125,870,736 inhabitants in 2030. Thus, several assumptions were used in the assessment of wheat production-consumption gap in 2030. The first assumption indicated that wheat breeding programs in Egypt will be successful in producing new wheat culti-vars tolerant to heat and water stresses with high water use efficiency, which will diminish wheat yield losses in 2030. This assumption is supported by the ongoing breeding programs implemented and the new wheat cultivars released by the Agricultural Research Center in Egypt.

The second assumption indicated that wheat consumption in 2013 namely; 218.6 kg/capita/year will be reduced to 175.0 kg/capita/year. This assumption is supported by the results of recent research on introduction of two new crops in Egypt, where its flour can be mix with wheat flour to make breads. These two crops are quinoa and cassava. Regarding to quinoa, it recently attracted attention because of its high nutritional value and strong growth potential under the extreme harsh conditions of drought and soil salinity (Shams 2012). The quinoa flour can substitute wheat flour with a ratio of 40 % breads industry (Shams 2011). Quinoa has the potential to become an important industrial and food crop in Egyptian farming systems of the twenty first century, because of its high ability to produce high-protein grain under the adverse growth conditions in reclaimed desert land (http://egypten.um.dk/en/about-us/news/newsdisplaypage/?newsid = 500ad6da-4da4-485b-b42e-0f6dc4 c1d781).The crop has been selected by the FAO as one of the main crops to play a major role in assuring food security in the twenty-first century because of this high nutritional value and its extreme resistance to adverse climatic conditions (Shams 2012). With respect to cassava, it was grown in Egypt on soils of low fertility and on already depleted soil. The crop has low production cost and low labor require-ments (Shams 2011). Cassava flour can substitute wheat flour with a ratio of 30 % in breads making (Shams 2011).

The third assumption indicted that the amount of water assigned to irrigation in 2030 will be the same as the assigned amount in 2012/2013.

Table 4.14 Projected wheat cultivated area and production under climate change effect in 2030

	Old land		New land		Total	
	Area (ha)	Production (ton)	Area (ha)	Production (ton)	Area (ha)	Production (ton)
Lower Egypt	659,204	4,561,012	116,453	735,618	775,657	5,296,630
Middle Egypt	200,645	1,402,226	25,504	147,014	226,149	1,549,240
Upper Egypt	144,134	983,616	76,985	427,195	221,119	1,410,811
Total	1,003,983	6,946,855	218,941	1,309,827	1,222,924	8,256,681

Table 4.15 Projected Egyptian population, wheat consumption and gap between production and consumption in 2030

	Population in July 2030	Total wheat consumption (ton)	Wheat food gap (ton)
Lower Egypt	76,547,694	13,395,846	8,099,217
Middle Egypt	27,588,466	4,827,982	3,278,741
Upper Egypt	21,734,576	3,803,551	2,392,740
Total	125,870,736	22,027,379	13,770,698

Projected Wheat Gap Under Surface Irrigation in 2029/2030

In wheat growing season of 2029/2030 and under surface irrigation, the total culti-vated area will be reduced to 1,003,983 and 218,941 hectares in the old and new land, respectively with total production of 8,256,681 tons (Table 4.14).

In 2030, wheat total consumption could reach 22,027,379 tons, which will make wheat consumption-production gap to be 13,770,698 tons (Table 4.15). Thus, the gap will increase by 25 %. The increase in the gap was mainly a result of increase in wheat water requirements and reduction in its cultivated area.

Thus, continue using surface irrigation to grow wheat in 2030 will result in very large wheat consumption-production gap. Therefore, cultivation on raised beds or using sprinkler system for wheat irrigation, in addition to wheat intercropping with other crops must be implemented to reduce that gap.

Projected Wheat Gap Under Cultivation on Raised Beds in 2029/2030

As it was stated before, cultivation of wheat on raised beds reduced the applied water by 20 % and increase yield by 15 %, compare to irrigation with surface irriga-tion. The saved irrigation water can cultivate a total of 267,729 hectares, which produce 1,632,138 tons of wheat (Table 4.16).

Changing wheat cultivation to raised beds cultivation can increase total wheat production to 9,839,987 tons and reduce wheat gap to 12,187,392 tons (Table 4.17).

Projected Wheat Gap Using Sprinkler System in 2029/2030

Using sprinkler system to irrigate wheat could increase the production of old land to be 8,197,289 tons and allow to cultivate extra 334,661 hectares with wheat using the saved irrigation water as a result of using sprinkler to irrigate wheat (Table 4.18).

This procedure will increase wheat national production to be 12,301,388 tons. Furthermore, it will result in reduction of wheat production-consumption gap to be 9,725,990 tons (Table 4.19).

Table 4.16 Projected new cultivated area with wheat, its production and total cultivated area under raised beds cultivation in Egypt in 2029/2030

	New cultivated area (ha)	Production of new area (ton)	Total cultivated area (ha)
Lower Egypt	175,788	1,110,430	911,406
Middle Egypt	53,505	308,424	200,519
Upper Egypt	38,436	213,284	465,631
Total	267,729	1,632,138	1,577,555

Table 4.17 Projected total wheat production and expected wheat food gap under raised beds cultivation in 2029/2030

	Total wheat production (ton)	Wheat food gap (ton)
Lower Egypt	6,574,535	6,821,311
Middle Egypt	1,920,992	2,906,989
Upper Egypt	1,344,459	2,459,092
Total	9,839,987	12,187,392

Table 4.18 Potential wheat cultivated area under sprinkler irrigation in the old and new lands of Egypt in 2029/2030

	Production of old land (ton)	New cultivated area (ha)	Production of new area (ton)	Total cultivated area (ha)
Lower Egypt	5,381,994	219,735	1,793,998	995,391
Middle Egypt	1,654,627	66,882	551,542	293,031
Upper Egypt	1,160,667	48,045	386,889	269,164
Total	8,197,289	334,661	2,732,430	1,557,585

Wheat Intercropping with Other Crops

Implementing the five suggested intercropping wheat systems will result in an increase in wheat total production by 1,451,060 tons (Table 4.20).

Thus, added wheat total production from intercropping to wheat production resulted from wheat cultivation on raised beds nationally; wheat gap will be reduced to 10,979,029 tons (Table 4.21).

Furthermore, if we assumed that all national wheat cultivated area will be irrigated with sprinkler and we will implement intercropping, food gap will be reduced to be 8,517,627 tons (Table 4.22).

Table 4.19 Wheat total production and wheat food gap under sprinkler system in 2029/2030

	Total production (ton)	Wheat food gap (ton)
Lower Egypt	7,973,454	5,422,393
Middle Egypt	2,353,183	2,474,798
Upper Egypt	1,974,751	1,828,800
Total	12,301,388	9,725,990

Table 4.20 Project wheat production under different systems of intercropping in 2029/2030

	Wheat potential production (ton) under intercropping with					
	Tomato	Cotton	Sugar beet	Sugarcane	Fruit trees	Total (ton)
Lower Egypt	91,314	564,025	218,423	–	183,930	1,057,692
Middle Egypt	99,037	43,910	62,348	–	26,862	232,158
Upper Egypt	58,391	8101	5971	56,842	31,905	161,211
Total	248,742	616,037	286,742	56,842	242,697	1,451,060

Table 4.21 Projected wheat total production and wheat food gap under intercropping and raised beds cultivation in 2029/2030

	Wheat total production (ton)	Wheat food gap (ton)
Lower Egypt	7,448,297	5,947,550
Middle Egypt	2,126,289	2,701,693
Upper Egypt	1,473,765	2,329,786
Total	11,048,350	10,979,029

Table 4.22 Wheat total production and wheat food gap under intercropping and sprinkler system in 2029/2030

	Wheat total production (ton)	Wheat food gap (ton)
Lower Egypt	8,847,215	4,548,631
Middle Egypt	2,558,480	2,269,502
Upper Egypt	2,104,057	1,699,494
Total	13,509,752	8,517,627

Conclusion

Wheat national production-consumption gap is a major economy dilemma in Egypt, where 49 % of wheat national consumption was imported in 2013. Furthermore, wheat consumption per inhabitant per year is also on the raise and it is the highest in the world. This rate is increasing as a result of population growth and it is expected to increase in the future in response to climate change. Thus, we tested several opportunities to increase wheat national production using the same currently applied amount of water and produce more wheat grains with it. Our assessment indicated that changing wheat cultivation from basins to raised beds can reduce wheat production-consumption gap to 32 %, as a result of cultivation of more lands with the saved irrigation water. Furthermore, if we change irrigation system nationally from surface to sprinkler, wheat gap will be 28 %. We also assessed the effect of intercropping wheat with other crops to increase its production nationally. Intercropping wheat in five intercropping systems, with tomato, sugar beet, cotton, sugarcane and under fruit trees can add an extra 1,667,283 tons of wheat. Adding this amount with the production under raised beds or irrigation with sprinkler, the gap will be lower, i.e. 23 or 19 %, respectively.

Under climate change in 2030, where the cultivated crops will consumed more irrigation water, the cultivated area of wheat will be lower and population will reach 125,870,736 inhabitants. Consequently, the gap will be larger and it will reach 63 %. Cultivation of wheat on raised beds or irrigating wheat with sprinkler will reduce the gap to be 55 or 44 %, respectively. Implementing intercropping with raised beds or irrigation with sprinkler will reduce the gap to be 49 or 38 %, respectively. Thus, opportunities exist to solve wheat production-consumption gap and we should take advantage of it.

References

Abd El-Zaher Sh, R., Shams, A. S., & Mergheny, M. M. (2013). Effect of intercropping pattern and nitrogen fertilization on intercropping wheat with tomato. *Egyptian Journal of Applied Sciences, 28*(9), 474–489.

Abdrabbo, M. (2011). Water management for some field crops grown under climate change conditions: A project report. Scientific Technology Development Fund. Egypt.

Abdrabbo, M., Ouda, S., & Noreldin, T. (2013). Modeling the effect of irrigation scheduling on wheat under climate change conditions. *Nature and Science Journal, 115*, 10–18.

Aboelenein, R., Sherif, M., Karrou, M., Oweis, T., Benli, B., & Farahani, H. (2011). Towards sustainable and improved water productivity in the old lands of Nile Delta. In *Water benchmarks of CWANA – Improving water and land productivities in irrigated systems, ICARDA*.

Abou-Elela, A. M. (2012). Effect of intercropping system and sowing dates of wheat intercropping with sugar beet. *Journal of Plant Production., 3*(12), 3101–3116.

Abouelenein, R., Oweis, T., El Sherif, M., Awad, H., Foaad, F., Abd El Hafez, S., Hammam, A., Karajeh, F., Karo, M., & Linda, A. (2009). Improving wheat water productivity under different methods of irrigation management and nitrogen fertilizer rates. *Egyptian Journal of Applied Science., 24*(12A), 417–431.

Aegic. (2015, April). *Global grain market series: Egypt.* Australian Export Grain Innovation Center, Australia.

Ahmed, A., Nagwa, M., Ahmed, R., Soha, R., & Khalil, A. (2013). Effect of intercropping wheat on productivity and quality of some promising sugarcane cultivars. *Minia Journal Agricultral Research Development, 33*(4), 557–583.

Andersen M. K. (2005). *Competition and complementarily in annual intercrops—The role of plant available nutrients.* Ph.D. thesis, Department of Soil Science, Royal Veterinary and Agricultural University, Copenhagen, Denmark. SamfundslitteraurGrafik, Frederiksberg, Copenhagen.

Bukhat N. M. (2005). *Studies in yield and yield associated traits of wheat Triticum aestivum L. genotypes under drought conditions.* MSc thesis, Department of Agronomy, Sindh Agriculture University, Tandojam, Pakistan.

Dencic, S., Kastori, R., Kobiljski, B., & Duggan, B. (2000). Evaporation of grain yield and its components in wheat cultivars and land races under near optimal and drought conditions. *Euphytica, 1*, 43–52.

Dunn, F., Williams, J., Verberg, K., & Keating, B. A. (1999). Can agricultural catchment emulate natural ecosystems in recharge control in southeastern Australia? *Agroforestry Systems, 45*, 343–364.

Eskandari, H., Ghanbari, A., & Javanmard, A. (2009). Intercropping of cereals and legumes for forage production. *Notulae Scientia Biologicae, 1*, 07–13.

Fernandez-Munoz, R., Gonzalez-Fernandez, J. J., & Cuartero, J. (1995). Variability of pollen tolerance to low temperatures in tomato and related with species. *Journal of Horticultural Science, 70*, 41–49.

Francis, C. A. (1986). Future perspectives of multiple cropping. In C. A. Francis (Ed.), *Multiple cropping systems* (pp. 351–370). New York: Macmillan.

Gupta, N. K., Gupta, S., & Kumar, A. (2001). Effect of water stress on physiological attributes and their relationship with growth and yield in wheat cultivars at different growth stages. *Journal of Agronomy, 86*(143), 7–1439.

Ibrahim, M., Ouda, S., Taha, A., El Afandi, G., & Eid, S. M. (2012). Water management for wheat grown in sandy soil under climate change conditions. *Journal of Soil Science and Plant Nutrition, 122*, 195–210.

Kamel, A. S., El-Masry, M. E., & Khalil, H. E. (2010). Productive sustainable rice based rotations in saline-sodic soils in Egypt. *Egyptian Journal of Agronomy, 32*(1), 73–88.

Khalil, F. A., Farag, H., El Afandi, G., & Ouda, S. A. (2009, March). Vulnerability and adaptation of wheat to climate change in Middle Egypt. In 13th conference on water technology, Hurghada, Egypt.

Li, J., & Rao, M. (2003). Field evaluation of crop yield as affected by non uniformity of sprinkler-applied water and fertilizers. *Agricultural Water Management, 59*, 1–13.

Liua, H., & Kanga, Y. (2006). Effect of sprinkler irrigation on microclimate in the winter wheat field in the North China Plain. *Agricultural Water Management, 84*(1–2), 3–19.

Majeed, A., Muhmood, A., Niaz, A., Javid, S., Ahmad, Z. A., Shah, S. S. H., & Shah, A. H. (2015). Bed planting of wheat (Triticum aestivum L.) improves nitrogen use efficiency and grain yield compared to flat planting. *The Crop Journal, 3*, 118–124.

Mansour, S. (2012). *Global agriculture information system; Egypt: Wheat and corn production on the rise: Grain and feed annual*. Annual report.

McMaster, G. S. (1997). Phonology, development, and growth of wheat (Triticum aestivum L.) shoot apex: A review. *Advaces in Agron, 59*, 63–118.

Metz, H. C. (1990). *Egypt: A country study*. Washington, DC: GPO for the Library of Congress. USA.

Nazir, M. S., Jabbar, A., Ahmad, I., Nawaz, S., & Bhatti, I. H. (2002). Production potential and economics of intercropping in autumn-planted sugarcane. *International Journal Agricultral and Biology, 41*, 140–141.

Ouda, S., Sayed, M., El Afandi, G., Khalil, F. (2010). *Developing an adaptation strategy to reduce climate change risks on wheat grown in sandy soil in Egypt*. 10th International Conference on Development of Dry lands, 12–15 December, Cairo, Egypt.

Ouda, S., Noreldin, T., AbouElenin, R., & Abd, E.-B. H. (2012). Improved agricultural management practices reduced wheat vulnerably to climate change in salt affected soils. *Egypt Journal of Agricultral Research, 904*, 499–513.

Ouda, S. A., Noreldin, T., & Abd El-Latif, K. (2015). Water requirements for wheat and maize under climate change in North Nile Delta. Span. *Journal of Agricultural Research, 13*(1), 1–7.

Pressman, E., Bar-Tal, A., Shaked, R., & Rosenfeld, K. (1997). The development of tomato root system in relation to the carbohydrate status of the whole plant. *Annals of Botany, 80*, 533–538.

Shams, A. S. (2011). Combat degradation in rain fed areas by introducing new drought tolerant crops in Egypt. *International Journal of Water Resources and Arid Environments, 1*(5), 318–325.

Shams, A. S. (2012). Response of quinoa to nitrogen fertilizer rates under sandy soil conditions. In Proceeding of the 13th international conference on agronomy (pp. 195–205). Banha.

Slafer, G. A., & Satorre, E. H. (1999). *Wheat: Ecology and physiology of yield determination*. New York: Haworth Press Technology and Industrial. ISBN 1560228741.

Snyder, R. L., Orang, M., Bali, K., & Eching, S. (2004). *Basic irrigation scheduling BIS*. http://www.waterplan.water.ca.gov/landwateruse/wateruse/Ag/CUP/Californi/Climate_Data_010804.xls

Taha A. (2012). *Effect of climate change on maize and wheat grown under fertigation treatments in newly reclaimed soil*. PhD Thesis, Tanta University, Egypt.

WOCAT. (2016). *Mixed fruit tree orchard with intercropping of Esparcet and annual crops in Muminabad District Tajikistan: Bog Orchard based agroforestry established on the hill slopes of Muminabad*. WOCAT_QT_Summary-T_TAJ043en.pdf

Wollenweber, B., Porter, J. R., & Schellberg, J. (2003). Lack of interaction between extreme high-temperature events at vegetative and reproductive growth stages in wheat. *Journal of Agronomy and Crop Science, 189*, 142–150.

Yang, F., Jørgensen, A. D., Li, H., Søndergaard, I., Finnie, C., Svensson, B., Jiang, D., Wollenweber, B., & Jacobsen, S. (2011). Implications of high-temperature events and water deficits on protein profiles in wheat (Triticum aestivum L. cv. Vinjett) grain. *Proteomics, 11*, 1684–1695.

Zhang, L., van der Werf, W., Zhang, S., Li, B., & Spiertz, J. H. J. (2007). Growth, yield and quality of wheat and cotton in relay strip intercropping systems. *Field Crops Research, 103*(3), 178–188.
Zhang, L., van der Werf, W., Bastiaans, L., Zhang, S., Li, B., & Spiertz, J. H. J. (2008). Light interception and utilization in relay intercrops of wheat and cotton. *Field Crops Research, 107*(1), 29–42.
Zohry, A. A. (2005). Effect of relaying cotton on some crops under bio-mineral N fertilization rates on yield and yield components. *Annals of Agricultral Science, 431*, 89–103.

Chapter 5
Increasing Land and Water Productivities to Reduce Maize Food Gap

Abd El-Hafeez Zohry and Samiha A.H. Ouda

Introduction

Maize is alternately an important food and feed crop. It has become an ever more vital component of global food security due to genetic and management practice changes that have driven yield gains over the last century (Ignacio et al. 2014). Maize, in Egypt, is planted in May and harvested in August; however some farmers plant it twice a year (May and July). There are three main summer crops compete for the available cultivable area, namely cotton, maize and rice. The area and production of yellow maize represented about 15 % of the total maize area and production in 2010/11 and the rest was white maize. The yield of white maize in 2010/11 was slightly higher than the yellow maize (Mansour 2012).

In 2012, the overall maize productivity of the United States of America was 9.0 ton/ha, while in China and Argentina was 6.0 ton/ha (FAO 2012). In the same year, maize productivity was 7.0 ton/ha, as an average on all governorates of Egypt. The value was increased to 8.0 ton/ha in 2014. This improvement in the national maize production is due to breeding programs and improved management practices. Over 80 % of the local maize crop is utilized for animal feed (mostly consumed in farms) and the rest is used for food purposes (either milled to produce glucose and fructose or consumed fresh). Feed consumption is estimated to be 9.0 million tons in 2012/13 (Mansour 2012).

Similar to wheat, two obstacles are expected to confront maize production-consumption gap in Egypt: population growth and climate change effect on maize

A.E.-H. Zohry
Crops Intensifications Research Department, Field Crops Research Institute,
Crops Agricultural Research Center, Giza, Egypt

S.A.H. Ouda (✉)
Water Requirements and Field Irrigation Research Department, Soils, Water and Environment
Research Institute, Agricultural Research Center, Giza, Egypt
e-mail: samihaouda@yahoo.com

© Springer International Publishing AG 2017
S. Ouda et al., *Future of Food Gaps in Egypt*, SpringerBriefs in Agriculture,
DOI 10.1007/978-3-319-46942-3_5

Table 5.1 Maize data in the old cultivated area in the growing season of 2013 in Egypt

	Cultivated area (ha)	Productivity (ton/ha)	Production (ton)	Water requirement (m³/ha)	Total water requirements (m³)
Lower Egypt	379,611	8.66	3,286,518	10,829	4,110,867,381
Middle Egypt	265,709	7.38	1,959,907	11,807	3,137,123,538
Upper Egypt	162,325	7.45	1,209,537	12,967	2,104,808,764
Total	807,645		6,455,962		9,352,799,684

water requirements. As it was indicated in Chap.2, in 2030 Egyptian population is projected to be 125,870,736 inhabitants. Furthermore, water requirements for maize in 2040 are expected to increase by 13 % in the Nile Delta, 16 % in Middle and 17 % in Upper Egypt (Ouda et al. 2016). This assessment was done using AR4 climate change scenarios. Thus, it is expected that maize gap will be enlarged in 2030. Thus, improved water and crop managements are considered as main driving factors for closing yield gaps for maize (Mueller et al. 2012). Examples of these management practices are cultivation on raised beds, using drip system for maize irrigation and intercropping maize with other summer crops.

The objective of this chapter was to explore alternatives to be used to reduce the current maize production-consumption gap through increasing land and water productivities. Furthermore, in 2030 under climate change effects and population increase, assessment of the contribution of these alternatives in reducing maize production-consumption gap were also implemented.

Current Maize Production-Consumption Gap

Maize is cultivated in both old clay soil and sandy soil under irrigation. In the old clay soil, maize is cultivated on narrow furrows under surface irrigation with 60 % application efficiency. In addition, this system has high-labor requirements for irrigation and fertilizer (Limon-Ortega et al. 2002). Furthermore, maize plants lodging was reported under surface irrigation (Wang et al. 2009). Table 5.1 showed maize cultivated area (ha), productivity (ton/ha) and total production (ton) in 2013 growing season in the old land of Egypt (collected data from Ministry of Agriculture and Land Reclamation, Egypt). Maize water requirements (m³/ha) was estimated by BISm model (Snyder et al. 2004) under surface irrigation with 60 % application efficiency. It is shown from the table that the total cultivated area in the old land was 807,645 ha produced 6,455,962 tons and consumed 9,352,799,684 m³ of irrigation water, which make its water productivity (total produced yield divided by total applied water) equal to 0.69 kg/m³ and its land productivity (produced per one squared meter of area) equal to 0.80 kg/m².

Table 5.2 Maize data in the sandy soil in the growing season of 2013 in Egypt

	Cultivated area (ha)	Productivity (ton/ha)	Production (ton)	Water requirement (m³/ha)	Total water requirements (m³)
Lower Egypt	62,333	8.35	520,527	7219	450,005,352
Middle Egypt	9352	6.32	59,087	7871	73,611,027
Upper Egypt	10,939	5.38	58,834	8644	94,563,019
Total	82,624	6.68	638,448		618,179,397

Table 5.3 Maize data (total) in the old and sand soils in the growing season of 2013 in Egypt

	Total cultivated area (ha)	Total production (ton)	Total water requirements (m³)
Lower Egypt	441,944	3,807,045	4,560,872,733
Middle Egypt	275,061	2,018,994	3,210,734,565
Upper Egypt	173,264	1,268,371	2,199,371,782
Total	890,268	7,094,410	9,970,979,080

Table 5.4 Egyptian population, maize consumption, new cultivated area to fill the gap and total required irrigation water

	Population in January 2014	Total maize consumption (ton)	Maize food gap (ton)	New area to fill the gap (ha)	Total required water (m³)
Lower Egypt	52,487,733	8,186,512	−4,379,467	524,436	3,786,132,310
Middle Egypt	18,672,890	2,912,411	−893,417	141,407	1,113,025,148
Upper Egypt	14,622,342	2,280,647	−1,012,276	188,215	1,627,015,740
Total	85,782,965	13,379,569	−6,285,159	854,058	6,526,173,198

Maize cultivated in the sandy soil of Egypt in 2013 growing season is irrigated with drip system, with 90 % application efficiency. Table 5.2 indicated that maize productivity and the applied water per hectare were lower than its counterpart in the old land. Thus, water productivity was higher, i.e. 0.98 kg/m³ and land productivity equal to 0.77 kg/m².

Table 5.3 revealed that total maize production in the growing season of 2013 was 7,094,410 tons and its total water requirements were 9,970,979,080 m³ (Table 5.3), which resulted in an overall water productivity equal 0.71 kg/m³ and land productivity equal to 0.80 kg/m².

Table 5.4 revealed that the highest maize production-consumption gap existed in Lower Egypt, followed by Upper Egypt. To fill this gap in Lower Egypt, 524,436 ha need to be cultivated. In Middle and Upper Egypt, 141,407 and 188,215 ha need to be cultivated, respectively. This large area requires 6,526,173,198 m³ of irrigation water under drip system. Due to limitation in our water resources, it will be difficult to acquire such large amount.

Table 5.5 Potential maize production under raised beds cultivation in the old land of Egypt

	Maize production (ton)	Water requirement (m³/ha)	Total water requirements (m³)
Lower Egypt	3,779,496	8663	3,288,693,905
Middle Egypt	2,253,893	9445	2,509,698,831
Upper Egypt	1,390,968	10,373	1,683,847,011
Total	7,424,356		7,482,239,747

Suggestions to Increase Maize Cultivated Area and Production

Use of Raised Beds Cultivation

Previous studies on raised beds cultivation have shown that it reduces seed mortality rates; increases water and nitrogen use efficiency, and improves soil quality. In addition, less labor is required for irrigation and fertilizer and it is better managed relative to conventional flat planting (Limon-Ortega et al. 2002). It can reduce crop lodging (Wang et al. 2009) and it exert control on applied water resulting in less water and nutrient loss through deep percolation, and reduced total water requirements (Dogan and Kirnak 2010). Raised beds cultivation could reduce applied water by 20 % and productivity can increase by 15 % (Abouelenein et al. 2009). Raised beds cultivation significantly and substantially increased maize growth, microbial functional groups and enzyme activities compare to flat planting, thus it increasing availability of essential crop nutrients by stimulating microbial activity (Zhang et al. 2012). The results in Table 5.5 revealed that using raised beds for maize cultivation can increase its productivity per hectare and its total production in the old land of Egypt. Furthermore, total water requirements were reduced and water productivity was increased to be 0.99 kg/m³and land productivity was increased to be 0.92 kg/m³.

Under these circumstances, the saved irrigation water can be used to cultivate new area with maize under drip system equal to 242,293 ha, which produces 1,716,556 tons of maize (Table 5.6). Thus, total maize production can reach 9,779,360 tons with the same applied irrigation water presented in Table 5.3. This amount of total maize production resulted from old cultivated land with 15 % increase in productivity under raised beds cultivation, production from added new area cultivated with maize as a result of irrigation water availability and production from sandy soil (Table 5.6). Thus, the overall water productivity can reach 0.98 kg/m³ under cultivation on raised beds and land productivity increased to 0.86 kg/m³.

Cultivation of maize on raised beds can reduce maize production-consumption gap to 3,600,209 tons. To fill this gap, cultivation of 472,466 ha on national level is needed with total required water equal to 3,571,564,013 m³ (Table 5.7).

Table 5.6 Potential new cultivated area with maize, productivity and total production under changing surface irrigation to raised beds cultivation in Egypt

	New cultivated area (ha)	Production of new area (ton)	Total cultivated area (ha)	Total maize production (ton)
Lower Egypt	113,883	951,018	555,827	5,251,041
Middle Egypt	79,713	503,629	354,773	2,816,609
Upper Egypt	48,697	261,909	221,961	1,711,710
Total	242,293	1,716,556	1,132,562	9,779,360

Table 5.7 Maize food gap, new cultivated area to fill the gap and total required irrigation water under raised beds cultivation

	Maize food gap (ton)	New area to fill the gap (ha)	Total required water (m³)
Lower Egypt	−2,935,471	351,519	2,537,770,246
Middle Egypt	−95,802	15,163	119,350,421
Upper Egypt	−568,937	105,784	914,443,346
Total	−3,600,209	472,466	3,571,564,013

Using Drip System to Irrigation Maize in the Old Land

The other option that can be used to reduce the applied water to maize is irrigation with drip system, where application efficiency increases from 60 to 90 % and maize productivity will increase by 18 % (Taha 2012). Some advantages of drip irrigation over other irrigation methods include improved water and nutrient management, improved saline water management, potential for improved yields and crop quality, reducing the incidence of diseases and weeds in dry row middles, greater control on applied water resulting in less water and nutrient loss through deep percolation, and reduced total water requirements (Dogan and Kirnak 2010). Furthermore, drip irrigation is an efficient method for minimizing water used in irrigation and can result in water saving if the correct management procedures were applied (Wang et al. 2006).

Using drip system to irrigate all the cultivated area of maize can increase its total production to 7,618,035 tons and it could add 403,822 ha to maize cultivated area as a result availability of irrigation water. These new lands are sandy, thus its productivity will be less than the productivity of the old land. Under this option, maize total production will result from old land irrigated with drip system with higher productivity per hectare, sandy soil with its same productivity and the productivity of the new added area with the same productivity of sandy soil presented in Table 5.2. Thus, total maize production will be 11,117,410 tons (Table 5.8). Furthermore, the overall water productivity can increase to be 1.11 kg/m³ and land productivity equal to 0.86 kg/m².

Using the above assumption, i.e. all maize cultivated area will be under drip system and that will result in higher maize production in Middle Egypt with an extra

Table 5.8 Potential maize production under drip system in the old and new lands of Egypt

	Old land production (ton)	New cultivated area (ha)	Production of new area (ton)	Total cultivated area (ha)	Total production (ton)
Lower Egypt	3,878,091	189,806	1,585,031	631,749	5,983,649
Middle Egypt	2,312,690	132,854	839,382	407,915	3,211,159
Upper Egypt	1,427,254	81,162	436,514	254,426	1,922,602
Total	7,618,035	403,822	2,860,927	1,294,091	11,117,410

Table 5.9 Maize food gap, new cultivated area to fill the gap and total required irrigation water under drip system

	Maize food gap (ton)	Modified maize gap (ton)	New area to fill the gap (ha)	Total required water (m³)
Lower Egypt	−2,099,681	−1,756,500	210,339	1,518,527,965
Middle Egypt	+343,180	0	0	0
Upper Egypt	−325,672	−325,672	60,553	523,448,444
Total	−2,082,173	−2,082,173	270,892	2,041,976,409

343,180 tons of maize and it can be used to reduce the production-consumption gap in Lower Egypt from 2,099,681 tons to be 1,756,500 tons. Furthermore, to fill the gap completely, cultivation of 210,339 and 60,553 ha is required in Lower and Upper Egypt, respectively and that will require the application of total 2,041,976,409 m³ of irrigation water (Table 5.9).

Intercropping Maize with Other Crops

Intercropping is a type of mixed cropping and defined as the agricultural practice of cultivating two or more crops in the same space at the same time (Hugar and Palled 2008). The important reason to grow two or more crops together is the increase in productivity per unit area of land in intercropping system (Reddy and Reddi 2007). Maize, have been recognized as a common component in most intercropping systems (Maluleke et al. 2005). Several scientists have worked, internationally, on cereal based intercropping, such as maize-potato, maize-soybean or maize-peanut, amongst many others (Mutsaers et al. 1993; Jiao et al. 2008 and Ijoyah and Fanen 2012). Intercropping maize with other summer crops is common in Egypt and it could be another option to increase maize productivity per hectare and save some of the applied water (Zohry and Ouda 2015). Four successful intercropping systems with maize were implemented in old clay soil in Egypt, namely tomato intercropping with maize, soybean or cowpea intercropping with maize and relay intercropping potato with maize. All these intercropping system are implemented in Egypt using raised beds cultivation. Another successful intercropping system can be

Fig. 5.1 Maize intercropped with tomato

implemented in sandy soil, which is intercropping maize with peanut, in addition to relay intercropping potato with maize, both irrigated with drip system.

Maize intercropping with tomato is very successful system in Egypt (Fig. 5.1), which is highly increases farmer's income. This system proved to reduced pests and diseases that usually exist in tomato monoculture (Pino et al. 1994). Hao (2013) indicated that maize intercropping with tomato has control effect on powdery mildew that occurs in tomato plants. Furthermore, spatial arrangement and different pattern of roots exploit soil nutrients in this system minimize plants competition (Ijoyah and Fanen 2012).

In Egypt, tomato is transplanted in April, 35–40 days before maize cultivation on one side of raised beds, with 100 % of its recommended planting density. Maize planting density is 70 % of the recommended and cultivated on the other side of the raised beds. The benefits of using this system are maize plants perform as sheds over tomato plants and protect its fruits from damage by sun rays. This system saved 100 % of the applied water for maize as it use the applied water to tomato plants (wafaa et al. 2013). Using this system, maize cultivated area can be increase by a percentage of summer tomato cultivated area.

The other effective intercropping system with maize is soybean or cowpea intercropped with maize. In case of soybean intercropping with maize (Fig. 5.2a), soybean is cultivated in first of May on the top of the raised beds in four rows and maize is cultivated on the both edges of the raised beds after 21 days from soybean planting. Planting densities are 50 and 80 % of the recommended densities for soybean and maize, respectively. Maize final yield is 90 % and soybean final yield is 60 % of its sole planting, respectively. In this system, maize plants take its water requirements from the applied water to soybean (Sherif and Gendy 2012).

a)   Maize intercropped with soybean   b)   Maize intercropped with cowpea

Fig. 5.2 Maize intercropped with soybean (**a**) and maize intercropped with cowpea (**b**)

Regarding to cowpea intercropping with maize (Fig. 5.2b), both crops cultivated on the same day. Maize is planted on both sides of the raised beds and cowpea is cultivated on the top of the raised beds. Planting density is 100 and 50 % of the recommended densities of maize and cowpea, respectively. Furthermore, in this system, cowpea reduces associated weeds (Hamd-Alla et al. 2014). Crop/weed competition is determined by growth habit of crop (Dimitrios et al. 2010), thus it is highly reduced under cowpea/maize intercropping system, where increased ground cover in intercropping system helps to reduce weed populations once the crops are established (Beets 1990). Maize productivity increases by 10 % with no reduction in cowpea productivity occurred under this system (Hamd-Alla et al. 2014).

Although the cultivated area of both soybean and cowpea are low in Egypt, maize yield increases, as a result of legume/cereal intercropping system can increase soil fertility via raising its organic content and available nitrogen fixed by legume (Singh et al. 1986), which reduce fertilizer requirements for cereal crops, reduces costly inputs and insure agricultural sustainability (Megawer et al. 2010). Legumes crops can improve soil quality, porosity, and structure (Rochester et al. 2001; McCallum et al. 2004) and influence specific microorganism populations in the rhizosphere (Kirkegaard et al. 2008; Osborne et al. 2010). Furthermore, legume crops facilitate the absorption of P and K in the soil by cereal crops, in addition to its role in providing N through N-fixing rhizobium (Bado et al. 2006). Kwari (2005) indicated that for optimum legume root development and rhizobia population in the soil, P is required, thus low absorptions of P by cereal crops, compared to legume crops provide it with the required amounts. Ferguson et al. (2013) indicated that legumes have the ability to remove calcium and magnesium in the soil more than cereals and replace it with hydrogen, which results in removing OH^- ions and increases H^+ thus lowering the soil pH and increase available K in the soil. Yan et al.

Fig. 5.3 Relay intercropping potato with maize

(1996) indicated that soil pH was significantly decreased after cultivation of field beans from 6.00 to 5.64.

Relay intercropping potato on maize is another successful system (Fig. 5.3) and highly profitable to the farmers. This system allows early cultivation for winter potato, where maize plants provide warmth to potato plants. Furthermore, it allowed early harvest for potato when its price is high. In this system, maize is cultivated in June with 100 % of its recommended planting density on one side of narrow furrow (60 cm in width), and potato is cultivated on the other side in September with 100 % planting density. Under this system, potato plants share its first and second irrigations with maize, thus this amount can be saved (Ibrahim 2006).

In sandy soil in Egypt, maize is intercropped with peanut (Fig. 5.4). Under this system, peanut is cultivated between the last week of April until mid of May and maize is planted 2–3 weeks after peanut cultivation under drip irrigation. Peanut planting density is 100 % and will attain 100 % of its yield, whereas maize planting density is 30 % and it will produce 40 % of its yield (Abd El-Zaher et al. 2007). This increase in maize productivity is due to more exposure to light. Maize intercropping with peanut enhanced the utilization efficiency of strong light by maize and that of weak light by peanut, making this intercropping system present an obvious yield advantage (Jiao et al. 2008). Furthermore, Inal et al. (2007) indicated that intercropping maize with peanut result in improving peanut seeds quality through reducing Fe deficiency stress and also contributes to better nutrition of plants increasing the availability of Zn, P and K, by affecting biological and chemical process in the rhizosphere.

Fig. 5.4 Maize intercropping with peanut

Maize New Added Area and Production Under Intercropping

We assumed that 90 % of the cultivated area with summer tomato will be assigned to be intercropped with maize. Furthermore, maize plants will produce 70 % of its yield under sole planting. Regarding to soybean intercropping with maize, we assumed it will be done on 100 % of the cultivated area of soybean and maize yield will increase by 10 %. Furthermore, the save irrigation water as a result of implementing this system can be used to cultivate new lands with maize under drip system. With respect to cowpea, it was only cultivated in Upper Egypt and we assumed that all its cultivated area will be intercropped with maize, which will result in 10 % increase in maize yield, however extra 208 m³/ha will be required to fulfill maize water requirements. Relay intercropping potato with maize will result in saving the first and second irrigation applied to potato. This amount of saved irrigation water in Upper Egypt can compensate for the needed water amount for maize under cow-pea/maize intercropping system. Moreover, the saved amount from Lower and middle Egypt will be used to cultivate more areas with maize under drip system. Finally, maize intercropping on peanut will be on 30 % of the cultivated area of peanut and it will produce 40 % of maize yield under sole planting. Table 5.10 indicated that using the five intercropping systems of maize can increase maize cultivated area by 79,887 ha.

This potential new cultivated area can produce 437,345 ton of maize, which can contribute in reducing maize production-consumption gap (Table 5.11).

Thus, if we implemented the suggested maize intercropping systems and cultivate maize on raised beds, Middle Egypt will develop self-sufficiency in maize and the gap will be reduced to 2,982,876 tons. To fill this gap, we need to cultivate

Table 5.10 Potential maize cultivated area under different systems of intercropping

	Potential maize cultivated area (ha) under intercropping with					
	Tomato	Soybean	Cowpea	Potato	Peanut	Total (ha)
Lower Egypt	23,792	113	–	9323	13,787	49,729
Middle Egypt	9086	8564	–	2456	2874	23,981
Upper Egypt	2670	517	1829	163	1149	6177
Total	35,547	9193	1829	11,941	17,810	79,887

Table 5.11 Potential maize production under different systems of intercropping

	Maize potential production (ton) under intercropping with					
	Tomato	Soybean	Cowpea	Potato	Peanut	Total (ton)
Lower Egypt	165,811	1262	–	22,646	56,338	246,058
Middle Egypt	53,948	81,898	–	4337	10,008	150,190
Upper Egypt	16,018	4937	16,084	17	4042	41,098
Total	235,777	88,097	16,084	27,000	70,387	437,346

Table 5.12 Maize food gap, new cultivated area to fill the gap and total required irrigation water under intercropping and raised beds cultivation

	Maize total production (ton)	Maize food gap (ton)	New area to fill the gap (ha)	Total required water (m³)
Lower Egypt	5,497,099	2,487,410	297,865	2,150,413,097
Middle Egypt	2,966,799	0	0	0
Upper Egypt	1,752,808	495,466	92,123	796,355,553
Total	10,216,706	2,982,876	389,988	2,946,768,649

Table 5.13 Maize food gap, new cultivated area to fill the gap and total required irrigation water under intercropping and drip system cultivation

	Maize total production (ton)	Maize food gap (ton)	New area to fill the gap (ha)	Total required water (m³)
Lower Egypt	6,229,706	1,360,252	162,889	1,175,963,895
Middle Egypt	3,361,349	0	0	0
Upper Egypt	1,963,700	284,574	52,912	457,392,256
Total	11,554,756	1,644,827	215,800	1,633,356,151

389,988 ha with maize under drip system. Furthermore, the total required water to cultivate this area will be 2,946,768,649 m³ (Table 5.12).

Thus, if we implemented the suggested maize intercropping systems and cultivate maize on drip system, Middle Egypt will develop self-sufficiency in maize and the gap will be reduced to 1,644,827 tons. To fill this gap, we need to cultivate 215,800 ha with maize under drip system. Furthermore, the total required water to cultivate this area will be 1,633,356,151 m³ (Table 5.13).

Effect of Climate Change on Maize Production

Climate change poses unprecedented challenges to agriculture because of the sensitivity of agricultural productivity to changing climate conditions. The abiotic stress that climate change will cause, i.e. water and heat stress can disturb physical and chemical processes in maize. Drought and temperature extremes can cause extensive economic loss to agriculture (Peng et al. 2004). The total yield loss depends on when the stress occurs (growth stage), as well as the duration and the severity of the stress. Early season drought reduces plant growth and inhibits plant development (Heiniger 2001). Drought occurring between 2 weeks before and 2 weeks after silking stage can cause significant reductions in kernel set and kernel weight (Schussler and Westgate 1991), resulting in an average of 20 % to 50 % yield loss (Nielsen 2007). High temperature stress at critical developmental stages of maize plants also causes significant yield loss (Lobell et al. 2011). Maize plants become susceptible to high temperatures after reaching eight-leaf stage to seventeen-leaf stages (Chen et al. 2010), which has significant impact on plant growth, architecture, ear size and kernel numbers (Farré and Faci 2006).

Assessment of Maize Production-Consumption Gap in 2030

To meet the demand of Egyptian population, which is projected to be 125,870,736 inhabitants in 2030, maize consumption per inhabitant per year should be reduced from 155.97 kg to 120.00 kg. This can be done by expending the area cultivated cassava, where it grown in Egypt on soils with low fertility and on already depleted soil. Shams (2011) indicated that cassava is a low cost production and low labor requirements. It can substitute maize starch with a ratio of 20–25 % and it can be used in animals feed. Furthermore, breeding for maize new tolerant hybrids to heat and water stress with high water use efficiency is going on now to help in reducing the vulnerability of maize to climate change. Furthermore, we did our assessment with the assumption that the amount of water assigned to irrigation in 2030 will be the same as the assigned amount in 2013.

Projected Maize Gap Under Surface Irrigation in 2030

In growing season of maize in 2030 and under surface irrigation, its total cultivated area will be reduced to 716,861 and 74,740 ha in the old and new land, respectively with total production of 6,328,950 tons (Table 5.14). Water productivity and land productivity will reduce to 0.63 kg/m^3 and 0.80 kg/m^2, respectively.

In 2030, total maize consumption could reach 15,104,488 tons, which make maize consumption-production gap 8,775,539 tons. To fill this gap, 1,214,871 ha

Table 5.14 Projected maize cultivated area and production under climate change effect in 2030

	Old land		New land		Total	
	area (ha)	Production (ton)	Area (ha)	Production (ton)	Area (ha)	Production (ton)
Lower Egypt	351,356	3,041,897	57,693	481,783	409,049	3,523,681
Middle Egypt	235,579	1,737,664	8292	52,387	243,870	1,790,050
Upper Egypt	129,926	968,127	8756	47,091	138,682	1,015,219
Total	716,861	5,747,688	74,740	581,261	791,602	6,328,950

Table 5.15 Projected Egyptian population, maize consumption and gap, new cultivated area to fill the gap and total required irrigation water in 2030

	Population in July 2030	Total maize consumption (ton)	Maize food gap (ton)	New area to fill the gap (ha)	Total required water (m³)
Lower Egypt	76,547,694	9,185,723	5,662,043	678,023	5,288,580,968
Middle Egypt	27,588,466	3,310,616	1,520,565	240,670	2,136,612,818
Upper Egypt	21,734,576	2,608,149	1,592,930	296,178	3,198,721,337
Total	125,870,736	15,104,488	8,775,539	1,214,871	10,623,915,123

will need to be cultivated and that will require the application of 10,623,915,123 m³ of irrigation water (Table 5.15).

Thus, using surface irrigation to grow maize in 2030 will result in large consumption-production gap. Therefore, cultivation on raised beds or using drip system for irrigation, in addition to maize intercropping with other crops must be implemented to reduce that gap.

Projected Maize Gap Under Cultivation on Raised Beds in 2030

Cultivation of maize on raised beds reduced the applied water by 20 % and increase yield by 15 %, compare to irrigation with surface irrigation. The saved irrigation water can cultivate a total of 247,293 ha, which produce 1,716,556 tons of maize and that will increase total maize production to 8,964,845 tons (Table 5.16). Water productivity can improve to 0.90 kg/m³ and land productivity will be 0.86 kg/m².

Changing maize cultivation to raised beds cultivation can reduce maize gap to 6,139,643 tons. To fill this gap, cultivation of 841,799 ha on the national level is needed with total required water equal to 6,465,699,131 m³ (Table 5.17).

Table 5.16 Projected new cultivated area with maize, its production and total production under raised beds cultivation in Egypt in 2030

	New cultivated area (ha)	Production of new area (ton)	Total cultivated area (ha)	Total maize production (ton)
Lower Egypt	113,883	951,018	527,572	4,969,727
Middle Egypt	79,713	503,629	324,643	2,561,029
Upper Egypt	48,697	261,909	189,563	1,434,089
Total	242,293	1,716,556	1,041,779	8,964,845

Table 5.17 Projected maize food gap, new cultivated area to fill the gap and total required irrigation water under raised beds cultivation in 2030

	Maize food gap (ton)	New area to fill the gap (ha)	Total required water (m³)
Lower Egypt	4,215,996	504,861	3,644,808,770
Middle Egypt	749,587	118,642	933,840,904
Upper Egypt	1,174,060	218,296	1,887,049,457
Total	6,139,643	841,799	6,465,699,131

Table 5.18 Potential maize production under drip system in the old and new lands of Egypt in 2030

	Production of old land (ton)	New cultivated area (ha)	Production of new area (ton)	Total cultivated area (ha)	Total production (ton)
Lower Egypt	351,356	175,678	1,467,054	584,727	5,577,020
Middle Egypt	235,579	117,789	744,200	361,660	2,853,730
Upper Egypt	129,926	64,963	349,391	203,646	1,550,615
Total	716,861	358,431	2,560,645	1,150,033	9,981,365

Projected Maize Gap Using Drip System in 2030

Using drip system to irrigate maize could increase the production of old land to be 716,861 ton and allow to cultivate extra 358,458 ha with maize using the saved irrigation water as a result of using drip to irrigate maize. This procedure will increase maize national production to be 9,981,365 tons (Table 5.18). Water productivity and land productivity will be 1.00 kg/m³ and 0.87 kg/m², respectively.

Furthermore, this procedure will result in reduction of maize production-consumption gap to be 5,123,123 tons. Cultivation of 526,010 ha is required with the application of 4,520,278,505 m³ of irrigation water will be required to fill that gap (Table 5.19).

Table 5.19 Maize food gap, new cultivated area to fill the gap and total required irrigation water under drip system in 2030

	Maize food gap (ton)	New area to fill the gap (ha)	Total required water (m³)
Lower Egypt	3,608,703	353,242	2,755,286,921
Middle Egypt	456,886	52,492	466,015,535
Upper Egypt	1,057,534	120,276	1,298,976,049
Total	5,123,123	526,010	4,520,278,505

Table 5.20 Projected maize production under different systems of intercropping in 2030

	Maize potential production (ton) under intercropping with					Total (ton)
	Tomato	Soybean	Potato	peanut	cowpea	
Lower Egypt	147,388	1134	14,972	46,948	–	210,442
Middle Egypt	47,954	73,506	2635	8340	–	132,434
Upper Egypt	14,238	4431	50	3368	16,084	38,171
Total	209,580	79,071	17,656	58,656	16,084	381,047

Table 5.21 Projected maize food gap, new cultivated area to fill the gap and total required irrigation water under intercropping and raised beds cultivation in 2030

	Maize Total production (ton)	Maize food gap (ton)	New area to fill the gap (ha)	Total required water (m³)
Lower Egypt	5,180,169	4,005,554	479,661	3,741,352,453
Middle Egypt	2,693,463	617,153	97,681	867,189,043
Upper Egypt	1,472,260	1,135,889	211,199	2,280,948,696
Total	9,345,892	5,758,597	788,540	6,889,490,192

Maize Intercropping with Other Crops

Intercropping maize with the suggested crops will result in an increase in its total production by 381,047 tons (Table 5.20).

Thus, added maize total production from intercropping to maize production resulted from maize cultivation on raised beds nationally; maize gap will be reduced to 5,758,597 tons. To fill this gap, we need to cultivate 788,540 ha with maize under drip system. Furthermore, the total required water to cultivate this area will be 6,889,490,192 m³ (Table 5.21).

Furthermore, if we assumed that all national maize cultivated area will be irrigated with drip and we will implement intercropping, maize food gap will be reduced to be 4,742,076 tons. A total new cultivated area of 647,824 ha is need to be cultivated to produce maize grains to fill the gap, which will require 5,676,973,394 m³ of irrigation water (Table 5.22).

Table 5.22 Maize food gap, new cultivated area to fill the gap and total required irrigation water under intercropping and drip system in 2030

	Maize total production (ton)	Maize food gap (ton)	New area to fill the gap (ha)	Total required water (m³)
Lower Egypt	5,787,462	3,398,261	406,938	3,174,115,933
Middle Egypt	2,986,164	324,452	51,353	455,902,163
Upper Egypt	1,588,786	1,019,363	189,533	2,046,955,297
Total	10,362,412	4,742,076	647,824	5,676,973,394

Conclusion

In 2013, maize production-consumption gap was 47 %, where maize was cultivated on narrow furrows. Thus, we assessed the effect of change cultivation methods for maize to raised beds. Our results showed that maize gap was reduced to 27 % as a result of reduction of the applied irrigation water and investment of the saved amounts in cultivation of new lands with maize on the national level. Similarly, our results also indicated that cultivating maize under drip system will result in reducing the gap to be 20 %. Furthermore, intercropping maize with soybean, cowpea, tomato, peanut and potato on national level could increase maize production. Thus, implementing maize intercropping systems and cultivation on raised beds can reduce the gap to be 23 %. Whereas, if we implement drip system on national level and use it to irrigate maize and also implement intercropping with other crops, the gap will be 17 % only.

Unfortunately, under climate change and population increase, maize gap will be 53, 41 and 38 % under narrow furrows, raised beds and drip system. Adding maize production under intercropping systems to production from raised beds cultivation or production from drip system, it will reduce the gap to be 38 and 35 %, respectively.

References

Abd El-Zaher, S., Mohamed, W., & Toaima, S. (2007). Intercropping maize with peanut under plant spacing and three planting dates. *45*(2), 545–560.

Abouelenein, R., Oweis, T., El Sherif, M., Awad, H., Foaad, F., Abd El Hafez, S., Hammam, A., Karajeh, F., Karo, M., & Linda, A. (2009). Improving wheat water productivity under different methods of irrigation management and nitrogen fertilizer rates. *Egyptian Journal of Applied Science, 24*(12A), 417–431.

Bado, B. V., Bationo, A., & Cescas, M. P. (2006). Assessment of cowpea and groundnut contributions to soil fertility and succeeding sorghum yields in the Guinean savannah zone of Burkina Faso (West Africa). *Biology and Fertility of Soils, 43*, 171–176.

Beets, W. C. (1990). *Raising and sustaining productivity of small holder systems in the tropics: A handbook of sustainable Agricultural development* (p. 40). Alkamaar: Agbe Publishing.

Chen, J., Xu, W., Burke, J. J., & Xin, Z. (2010). Role of phosphatidic acid in high temperature tolerance in maize. *Crop Science, 50*, 2506–2515.

Dimitrios, B., Panyiota, P., Aristidis, K., & Aspasia, E. (2010). Weed suppression effects of maize-vegetable in organic farming. *International Journal of Pest Management, 56*, 173–181.

Dogan, E., & Kirnak, H. (2010). Water temperature and system pressure effect on drip lateral properties. *Irrigation Science, 28*, 407–419.

FAO. (2012). FAOSTAT 2012: FAO Statistical database. FAO, Rome http://faostat.fao.org/site

Farré, I., & Faci, J. M. (2006). Comparative response of maize Zea mays L. and sorghum Sorghum bicolor L. Moench to deficit irrigation in a Mediterranean environment. *Agricultural Water Management, 83*, 135–143.

Ferguson, B. J., Lin, M. H., & Gresshoff, P. M. (2013). Regulation of legume nodulation by acidic growth conditions. *Plant Signal Behavior, 8*(3), e23426.

Hamd-Alla, W. A., Shalaby, E. M., Dawood, R. A., & Zohry, A. A. (2014). Effect of cowpea (Vigansinensis L.) with maize (Zea mays L.) intercropping on yield and its components. *International Scholarly and Scientific Research and Innovation, 8*(11), 1170–1176.

Hao, W. (2013). Control effect of tomato and maize intercropping against tomato powdery mildew. *Plant Diseases and Pests, 4*(2), 22–24.

Heiniger, R. W. (2001). The impact of early drought on corn yield. North Carolina State University. http://www.ces.ncsu.edu/plymouth/cropsci/docs/early_drought_impact on_corn.html

Hugar, H. Y., & Palled, Y. B. (2008). Studies on maize-vegetable intercropping systems. *Karnataka Journal Agricultural Sciences, 21*, 162–164.

Ibrahim, S. (2006). Intercropping potato with maize. *Annals of Agricultural Science, 44*(3), 799–817.

Ignacio, C., Roozeboom, K., & Jardine, D. (2014). *Effect of water-logged soils on corn growth and yield*. Kansas State University.

Ijoyah, M. O., & Fanen, F. T. (2012). Effects of different cropping pattern on performance of maize-soybean mixture in Makurdi. *Nigerian Scientific Journal of Crop Science, 1*(2), 39–47.

Inal, A., Gunes, A., Zhang, F., & Cakmak, I. (2007). Peanut/maize intercropping induced changes in rhizosphere and nutrient concentrations in shoots. *Plant Physiology and Biochemistry, 45*(5), 350–356.

Jiao, N. Y., Zhao, C., Ning, T. Y., & Chen, M. C. (2008). Effects of maize-peanut intercropping on economic yield and light response of photosynthesis. *Chinese Journal of Applied Science, 19*, 981–985.

Kirkegaard, J., Christen, O., Krupinsky, J., & Layzell, D. (2008). Review: Break crop benefits in temperate wheat production. *Field Crops Research, 107*, 185–195.

Kwari, J. D. (2005). Soil fertility status in some communities of southern Borno. Final report to PROSAB Project, Maiduguri, Nigeria.p. 21.

Limon-Ortega, A., Sayre, K. D., Drijber, R. A., & Francis, C. A. (2002). Soil attributes in a furrow-irrigated bed planting system in northwest Mexico. *Soil and Tillage Research, 63*, 123–132.

Lobell, D. B., Bänziger, M., Magorokosho, C., & Vivek, B. (2011). Nonlinear heat effects on African maize as evidenced by historical yield trials. *Nature Climate Change, 1*, 42–45.

Maluleke, M. H., Bediako, A. A., & Ayisi, K. K. (2005). Influence of maize-lablab intercropping on Lepidopterous stem borer infestation in maize. *Journal of Entomology, 98*, 384–388.

Mansour, S. (2012). Global agriculture information system; Egypt: Wheat and corn production on the rise: Grain and feed annual. Annual report.

McCallum, M. H., Kirkegaard, J. A., Green, T., Cresswell, H. P., Davies, S. L., & Angus, J. F. (2004). Improved subsoil macro-porosity following perennial pastures. *Australian Journal of Experimental Agriculture, 44*, 299–307.

Megawer, E. A., Sharaan, A. N., & El-Sherif, A. M. (2010). Effect of intercropping patterns on yield and its components of barley, lupine or chickpea grown in newly reclaimed soil. *Egyptian Journal of Applied Science, 25*(9), 437–452.

Mueller, N. D., Gerber, J. S., Johnston, M., Ray, D. K., Ramankutty, N., & Foley, J. A. (2012). Closing yield gaps through nutrient and water management. *Nature, 490*, 254–257.

Mutsaers, H. J. N., Ezuma, H. C., & Osiru, D. S. O. (1993). Cassava based intercropping: A review. *Field Crops Research, 34*, 431–457.

Nielsen, R. L. (2007). Assessing effects of drought on corn grain yield. West Lafayette, IN: Purdue University. http://www.kingcorn.org/news/articles.07/Drought-0705.html

Osborne, C. A., Peoples, M. B., & Janssen, P. H. (2010). Detection of a reproducible, single-member shift in soil bacterial communities exposed to low levels of hydrogen. *Applied Environmental Microbiology, 76*, 1471–1479.

Ouda, S., Abd El-Latif, K., & Khalil, F. (2016). Water requirements for major crops. In *Major crops and water scarcity in Egypt* (pp. 25–31). Springer Publishing House.

Peng, S., Huang, J., Sheehy, J. E., Laza, R. C., Visperas, R. M., Zhong, X., Centeno, G. S., Khush, G. S., & Cassman, K. G. (2004). Rice yields decline with higher night temperature from global warming. *National Academy of Sciences, 101*, 9971–9975.

Pino, M., De-Los, A., Bertoh, M., & Espinosa, R. (1994). Maize as a protective crop for tomato in conditions of environmental stress. *Cultural Tropical, 15*, 60–63.

Reddy, T. Y., & Reddi, G. H. S. (2007). *Principles of Agronomy* (pp. 468–489). Ludhiana: Kalyam Publishers.

Rochester, I. J., Peoples, M. B., Hulugalle, N. R., Gault, R. R., & Constable, G. A. (2001). Using legumes to enhance nitrogen fertility and improve soil conditions in cotton cropping systems. *Field Crops Research, 70*, 27–41.

Schussler, J. R., & Westgate, M. E. (1991). Maize kernel set at low water potential: II. Sensitivity to reduce assimilate supply at pollination. *Crop Science, 31*, 1196–1203.

Shams, A. S. (2011). Combat degradation in rain fed areas by introducing new drought tolerant crops in Egypt. *International Journal of Water Resources and Arid Environments, 1*(5), 318–325.

Sherif, S. A., & Gendy, E. K. (2012). Growing maize intercropped with soybean on beds. *Egyptian Journal of Applied Science, 27*(9), 409–423.

Singh, N. B., Singh, P. P., & Nair, K. P. (1986). Effect of legume intercropping on enrichment of soil nitrogen, bacterial activity and productivity of associated maize crops. *Experimental Agriculture, 22*, 339–344.

Snyder, R. L., Orang, M., Bali, K., & Eching, S. (2004). Basic irrigation scheduling BIS. http://www.waterplan.water.ca.gov/landwateruse/wateruse/Ag/CUP/Californi/Climate_Data_010804.xls

Taha, A. (2012). Effect of climate change on maize and wheat grown under fertigation treatments in newly reclaimed soil. PhD thesis, Tanta University, Egypt.

Wafaa, M., Ahmed, N. R., & Abd El-Hakim, W. M. (2013). Effect of intercropping dates of sowing and N fertilizers on growth and yield of maize and tomato. *Egyptian Journal of Applied Sciences, 28*(12B), 625–644.

Wang, F. X., Kang, Y., & Liu, S. P. (2006). Effects of drip irrigation frequency on soil wetting pattern and potato growth in North China plain. *Agricultural Water Management, 79*, 248–264.

Wang, F., He, Z., Sayre, K., Li, S., Si, J., Feng, B., & Kong, L. (2009). Wheat cropping systems and technologies in China. *Field Crops Research, 111*, 181–188.

Yan, F., Schubert, S., & Mengel, K. (1996). Soil pH changes during legume growth and application of plant material. *Biology and Fertility of Soils, 23*(3), 236–242.

Zhang, X., Ma, L., Gilliam, F. S., & Li, Q. W. C. (2012). Effects of raised-bed planting for enhanced summer maize yield on rhizosphere soil microbial functional groups and enzyme activity in Henan Province. *China Field Crops Research, 130*, 28–37.

Zohry, A., & Ouda, S. (2015). *Facing water scarcity in Egypt by intercropping: Maximizing maize production to reduce its gap under changing climate*. Saarbrücken: LAMBERT Academic Publishing.

Chapter 6
Solution for Faba Bean Production-Consumption Gap

Abd El-Hafeez Zohry and Samiha A.H. Ouda

Introduction

Faba bean is one of the main winter pulses seeds in Egypt. Faba bean has high nutrition value; therefore it is very poplar diet for the Egyptians. Faba bean seeds consumption was estimated by 5.90 kg/capita/year between 1984 and 1988. In 2013, its consumption increased to be 6.33 kg/capita/year. As a result of limited cultivated area in Egypt and expansion in sugar beet cultivation to reduce sugar production-consumption gap in Egypt, the cultivated area of faba bean was highly reduced. According to Ministry of Agriculture and Land Reclamation in Egypt, the national cultivated area of faba bean in 2012/2013 was only 41,975 ha, which produced 194,259 ton of faba bean seeds. Egypt is the world's largest importer of faba beans, where its annual requirement of it is 480,000–520,000 tons. These imports are roughly split between United Kingdom, France and Australia (Poutney 2014). In 2014, Egypt imported 390,000 ton of faba bean to fulfill the national consumption (Field Crops Research Institute; Agricultural Research Center; Egypt).

Faba bean production-consumption gap in Egypt was estimated to be 73 % in 2014. To reduce this gap, unconventional solutions need to be conducted. Faba bean is cultivated in the old land in the Nile Delta and Valley under surface irrigation with 60 % application efficiency. However, cultivation of faba bean on raised beds is

A.E.-H. Zohry (✉)
Crops Intensifications Research Department, Field Crops Research Institute,
Crops Agricultural Research Center, Giza, Egypt
e-mail: abdelhafeezzohry@yahoo.com

S.A.H. Ouda
Water Requirements and Field Irrigation Research Department, Soils, Water and Environment
Research Institute, Agricultural Research Center, Giza, Egypt
e-mail: samihaouda@yahoo.com

© Springer International Publishing AG 2017   75
S. Ouda et al., *Future of Food Gaps in Egypt*, SpringerBriefs in Agriculture,
DOI 10.1007/978-3-319-46942-3_6

common practice in many countries around the world, such as Australia. Dean and Mendham (2003) indicated that planting faba bean on raised beds with 30 plant/m^2 population density gave the highest yield. Moreover, in India, raised beds planting for faba bean proved superior over flat beds and phosphorus availability was significantly higher in raised bed planting (Singh et al. 2013). In Egypt, cultivation of faba bean on raised beds saved 20 % of the applied water under surface irrigation and its yield was increased by 15 % (http://www.icarda.org/update/producing-more-food-less-resources #sthash. ZYyk MnEI.dpbs).

Furthermore, changing surface irrigation to drip system can increase application efficiency to 90 % and save 30 % of the applied water to be used in cultivating new lands under drip system (Noreldin et al. 2015). Furthermore, irrigating faba bean with drip system can increase its yield by 11 %, compare to its yield under surface irrigation (Abouelenein et al. 2009). Drip irrigation system has a unique capacity to allow different plant distributions around the emitters and has many advantages in the cultivation of field row crops in arid and semiarid regions (Al-Suhaibani et al. 2013).

Another approach can be implemented to reduce this gap is intercropping faba bean with other winter crops, such as sugar beet and tomato. Furthermore, intercropping faba bean with sugarcane in South Egypt, as well as interplant it under young fruit trees can increase its national production and reduce its production-consumption gap.

Thus, the objective of this chapter was to test options to be used to reduce the current faba bean production-consumption gap. Furthermore, in 2030 under climate change effects and population increase, assessment of the contribution of these options in reducing the gap was also implemented.

Current Faba Bean Production-Consumption Gap

Faba bean is cultivated in old clay soils under surface system. Limon-Ortega et al. (2002) indicated that surface system is labor intensive with respect to irrigation and fertilizer. Furthermore, in sandy soils of Egypt faba bean is cultivated under drip system. Table 6.1 presents faba bean cultivated area in the old lands (ha), productivity (ton/ha) and total production (ton) in 2012/2013 growing season in Egypt (collected data from Ministry of Agriculture and Land Reclamation, Egypt). Moreover, faba bean water requirements (m^3/ha) was estimated by BISm model (Snyder et al. 2004) under surface irrigation with 60 % application efficiency. It is shown from the Table that the total cultivated area in the old land was only 28,032 ha, which produced 94,111 tons of faba bean seeds and consumed 189,587,053 m^3 of irrigation water.

With respect to faba bean cultivated in the sandy soil of Egypt in the same growing season under drip system, Table 6.2 indicated that the cultivated area was low,

Table 6.1 Faba bean data in the old cultivated area in the growing season of 2012/2013 in Egypt under surface irrigation

	Cultivated area (ha)	Productivity (ton/ha)	Production (ton)	Water requirement (m³/ha)	Total water requirements (m³)
Lower Egypt	23,845	3.45	82,157	6512	155,287,153
Middle Egypt	1278	2.45	3127	6760	8,641,300
Upper Egypt	2908	3.04	8827	8822	25,658,600
Total	28,032		94,111		189,587,053

Table 6.2 Faba bean data in the sandy soil in the growing season of 2013 in Egypt under drip system

	Cultivated area (ha)	Productivity (ton/ha)	Production (ton)	Water requirement (m³/ha)	Total water requirements (m³)
Lower Egypt	12,706	3.82	48,536	4341	55,162,318
Middle Egypt	25	2.79	71	4507	114,541
Upper Egypt	1212	2.58	3128	5882	7,129,023
Total	13,943		51,735		62,405,882

Table 6.3 Faba bean data (total) in the old and sand soils in the growing season of 2012/2013 in Egypt

	Total cultivated area (ha)	Total production (ton)	Total water requirements (m³)
Lower Egypt	36,551	130,693	210,449,471
Middle Egypt	1304	3198	8,755,842
Upper Egypt	4120	11,955	32,787,622
Total	41,975	145,846	251,992,935

i.e. 13,943 ha and it produced 51,735 tons of faba bean seeds using 62,405,882 m³ of irrigation water.

Table 6.3 revealed that total faba bean production in the growing season of 2012/2013 was 41,975 tons and its total water requirements were 251,992,935 m³ (Table 6.3).

Table 6.4 revealed that the highest faba bean production-consumption gap was existed in Lower Egypt, followed by Middle Egypt. To fill this gap, 51,667; 40,523 and 30,725 ha in Lower, Middle and Upper Egypt, respectively need to be cultivated, with a total of 122,914 ha. This large area requires 587,639,349 m³ to be irrigated under drip system. The existed limitation in our water resources will make it difficult to secure such large amount.

Table 6.4 Egyptian population, faba bean consumption, new cultivated area to fill the gap and total required irrigation water

	Population in august 2013	Total faba bean consumption (ton)	Faba bean food gap (ton)	New area to fill the gap (ha)	Total required water (m³)
Lower Egypt	51,826,181	328,060	197,366	51,667	224,310,345
Middle Egypt	18,388,014	116,396	113,198	40,523	182,617,849
Upper Egypt	14,414,787	91,246	79,291	30,725	180,711,155
Total	84,628,982	535,701	389,855	122,914	587,639,349

Table 6.5 Potential faba bean cultivated area, productivity and production under raised beds cultivation in the old land of Egypt

	Cultivated area (ha)	Productivity (ton/ha)	Total production (ton)	Water requirement (m³/ha)	Total water requirements (m³)
Lower Egypt	23,845	3.96	94,481	5210	124,229,723
Middle Egypt	1278	2.81	3596	5408	6,913,040
Upper Egypt	2908	3.49	10,151	7058	20,526,880
Total	28,032		108,228		151,669,643

Suggestions to Increase Faba Bean Cultivated Area and Production

Use of Raised Beds Cultivation

A feature of the raised beds planting system is the hole between the beds. This hole comprises the unplanted shoulder of a bed, the furrow itself and the shoulder of the adjacent bed. It increases the ability of the cultivar to capture the solar radiation falling in these holes between the beds (Fischer et al. 2005). Many researchers internationally reported that raised beds planting saved a reasonable amount of applied irrigation water could reach to an average of 40 %, as compare to flat planting (Kumar et al. 2010). Furthermore, it could increase legume crops yield by 17 % over flat planting technique (Pramanik et al. 2009). In Egypt, raised beds cultivation could reduce applied water by 20 % and productivity in tons per fully irrigated hectare can increase by 15 % (Abouelenein et al. 2009).

The results in Table 6.5 indicated that using raised beds for faba bean cultivation can increase its total production in the old land of Egypt to be 108,228 ton. Furthermore, total water requirements were reduced to be 151,669,643 m³.

The saved irrigation water amount as a result of cultivation on raised beds can be used to cultivate new area with faba bean equal to 8410 ha, which can produce 30,650 tons of faba bean seeds (Table 6.6). Thus, total faba bean production can reach 190,613 tons with the same applied irrigation water presented in Table 6.4. This amount of total faba bean production resulted from old cultivated land with 15 % increase in productivity under raised beds cultivation, production from added

Table 6.6 Potential new cultivated area with faba bean, productivity and total production under changing surface irrigation to raised beds cultivation in Egypt

	New cultivated area (ha)	Production of new area (ton)	Total cultivated area (ha)	Total faba bean production (ton)
Lower Egypt	7154	27,327	43,705	170,344
Middle Egypt	384	1071	1687	4738
Upper Egypt	873	2252	4993	15,531
Total	8410	30,650	50,385	190,613

Table 6.7 Faba bean food gap, new cultivated area to fill the gap and total required irrigation water under raised beds cultivation

	Faba bean food gap (ton)	New area to fill the gap (ha)	Total required water (m³)
Lower Egypt	157,716	41,287	179,246,989
Middle Egypt	111,658	39,971	180,132,890
Upper Egypt	75,715	29,339	172,561,793
Total	345,089	110,597	531,941,672

new area cultivated with faba bean as a result of irrigation water availability and production from new sandy soil (Table 6.6).

Cultivation of faba bean on raised beds can reduce its production-consumption gap to 345,089 ton. To fill this gap, cultivation of 110,597 ha on national level is needed with total required water equal to 531,941,672 m³ (Table 6.7).

Irrigation of Faba Bean in the Old Land with Drip System

The other option that can be used to reduce applied water to faba bean is irrigation with drip system, where application efficiency increases from 60 to 90 % and faba bean productivity will increase by 11 % (Abouelenein et al. 2009). Drip irrigation has advantages over conventional systems of irrigation to be an efficient means of applying irrigation water; especially where water amounts are limited. Thus, water could be saved and both crop quantity and quality could be increased (Mansour et al. 2014).

Using drip system to irrigate all the cultivated area of faba bean can increase its total production to 206,340 tons through adding 103,522 ha to faba bean cultivated area as a result availability of irrigation water. Under this option, faba bean total production will result from old land irrigated with drip system, sandy soil with its same productivity and the productivity of the new added area with the same productivity of sandy soil in Table 6.8.

Using the above assumption, i.e. all faba bean cultivated area will be under drip system will result in reduction of the production-consumption gap 143,606; 111,100 and 74,655 tons in Lower, Middle and Upper Egypt, respectively. Furthermore, to

Table 6.8 Potential faba bean production under drip system in the old and new lands of Egypt

	Old land production (ton)	New cultivated area (ha)	Production of new area (ton)	Total cultivated area (ha)	Total production (ton)
Lower Egypt	90,373	11,923	45,545	48,474	184,454
Middle Egypt	3440	639	1785	1943	5296
Upper Egypt	9710	1454	3753	5575	16,590
Total	103,522	14,016	51,083	55,991	206,340

Table 6.9 Faba bean food gap, new cultivated area to fill the gap and total required irrigation water under drip system

	Faba bean food gap (ton)	New area to fill the gap (ha)	Total required water (m³)
Lower Egypt	143,606	37,593	163,210,677
Middle Egypt	111,100	39,772	179,232,950
Upper Egypt	74,655	28,928	170,146,527
Total	329,361	106,293	512,590,154

fill the gap completely, cultivation of 37,593 in Lower Egypt is required, as well as 39,772 and 28,928 ha in Middle and Upper Egypt, respectively. Extra irrigation water amounts for 512,590,154 m^3 will be required under this option (Table 6.9).

Intercropping Faba Bean with Other Crops

Intercropping systems are generally recommended to get sTable yields. The total water used in intercropping system is almost the same as for sole crops, but yields are increased than sole crops (Singh et al. 2013). Intercropping better exploits environmental resources, thus improving soil fertility and increase crop yield (Addo-Quaye et al. 2011). Intercropping with legumes is an efficient cropping system in terms of resources utilization. For instance, crops with different roots traits can explore various organic P sources in P-deficient soils (Raghothama and Karthikeyan 2005). It also increased P mobilization via chelation of Ca^{2+} by citrate exuded from the roots of one of the intercropped species (Li et al. 2007). Furthermore, intercropping with legumes enhanced N acquisition significantly, compared with monoculture (Li et al. 2001). Because faba bean is a nitrogen-fixing plant, capable of fixing about 150–300 kg N/ha of atmospheric nitrogen (Singh and Umrao 2013), it is good candidate for implementing intercropping system.

There are four popular intercropping systems with faba bean in Egypt, namely faba bean intercropped with winter tomato, faba bean intercropped with sugar beet, faba bean intercropped with sugarcane in Upper Egypt and faba bean interplant under young fruit trees.

Fig. 6.1 Faba bean intercropped with tomato (**a**) and faba bean intercropped with sugar beet (**b**)

Tomato is one of the most important vegeTable crops grown in large areas in Egypt throughout the year. Several advantages of intercropping faba bean with tomato are obtained (Fig. 6.1a), namely protection of tomato plants from abiotic stress (high or low temperatures). Furthermore, this system can increase the cultivated area of faba bean and consequently increase its production. Under this system, tomato is cultivated in the beginning of September, with 100 % planting density and faba been is cultivated in November with 50 % of recommended planting density. Both crops are harvested in the same date, which increase tomato season length by 2 months. Thus, this system increases the farmer's profit (Ibrahim 2006).

Sugar beet is an important winter crop in Egypt and it contributes in reducing sugar production-consumption gap in Egypt. Its cultivation resulted in reduction of the assigned area to other winter crops, especially faba been. Thus, to solve the problem of limited land and water resources, faba bean is successfully intercropped with sugar beet (Fig. 6.1b). Faba bean intercropped with sugar beet increases land and water productivity because no extra irrigation water or fertilizer is applied to faba bean. Sugar beet is cultivated to maintain 100 % of its recommended planting density and faba bean is cultivated by 12.5 % of its recommended planting density. As a result, the farmer can obtain 100 and 25 % of sugar beet and faba bean, respectively (Abd El-Zaher and Gendy 2014).

Sugarcane is a high water-consuming crop, not only because it has long growing season, but also it has large above ground biomass. It is the main cultivated crop in south Egypt. Sugarcane offers a unique potential for intercropping. It is planted in wide rows (100 cm), and takes several months to develop its canopy, during which time the soil and solar energy goes to waste. The growth rate of sugarcane during its early growth stages is slow, with leaf canopy providing sufficient uncovered area for growing of another crop (Nazir et al. 2002). Thus, intercropping faba bean with sugarcane will not require the application of any extra irrigation water as it will use

Fig. 6.2 Faba bean intercropped with sugarcane (**a**) and faba bean interplaned under fruit trees (**b**)

the applied water to sugarcane to fulfill its required water. Furthermore, intercropping on sugarcane provide extra income for farmers during the early growth stage of sugarcane. In addition, intercropping faba bean with sugarcane reduces the applied nitrogen fertilizer to sugarcane in this growth stage. Under this system (Fig. 6.2a), faba bean is cultivated in October in two rows with 50 % of its recommended planting density and harvested in March (Farghly 1997).

Similarly, interplant faba bean under young evergreen fruit tree (1–3 years old) or deciduous fruit trees can increase the cultivated area with faba bean and consequently its national production (Fig. 6.2b). This practice gives an extra economic incentive and also improves land productivity. It also can be done under young evergreen fruit trees by separation between fruit trees and faba bean cultivated area to prevent the runoff of irrigation water to these trees (WOCAT 2016).

Faba Bean New Added Area and Production Under Intercropping

We assumed that 45 % of the cultivated area with winter tomato will be assigned to be intercropped with faba bean and faba bean will produce 60 % of its yield under sole planting. Regarding to intercropping faba bean with sugar beet, we assumed it will be done on 30 % of the cultivated area of sugarcane and faba bean will produce 25 % of its yield under sole planting. Regarding to intercropping faba bean on sugarcane and under fruit trees, we assumed that 17 % of its cultivated area will be used for intercropping on sugarcane. Furthermore, faba bean productivity will be 100 % under intercropping on both sugarcane and under fruit trees. Table 6.10 indicted that using these intercropping systems can increase faba bean cultivated area by 254,784 ha.

Table 6.10 Potential faba bean cultivated area under different systems of intercropping

	Potential faba bean cultivated area (ha) under intercropping with				
	Tomato	Sugar beet	Sugarcane	Fruit trees	Total (ha)
Lower Egypt	16,138	82,353	Not planted	42,196	140,687
Middle Egypt	34,658	23,273	Not planted	6101	64,033
Upper Egypt	20,926	2282	19,435	7421	50,064
Total	71,722	107,909	19,435	55,718	254,784

Table 6.11 Potential faba bean production under different systems of intercropping

	Faba bean potential production (ton) under intercropping with				
	Tomato	Sugar beet	Sugarcane	Fruit trees	Total (ton)
Lower Egypt	38,366	70,935	Not planted	145,382	254,683
Middle Egypt	58,498	14,233	Not planted	14,924	87,654
Upper Egypt	43,823	1732	58,987	160,306	264,848
Total	140,687	86,899	58,987	320,612	607,185

Table 6.12 Faba bean total production and its gap under intercropping and raised beds cultivation

	Faba bean total production (ton)	Faba bean food gap (ton)	Amount of extra faba bean (ton)
Lower Egypt	425,027	+96,967	96,967
Middle Egypt	92,392	−24,004	0
Upper Egypt	280,379	+189,133	156,129
Total	797,798		262,096

This potential new cultivated area with faba bean can produce 607,185 tons, which can contribute in reducing faba bean production-consumption gap (Table 6.11).

Thus, if we implemented the suggested faba bean intercropping systems and cultivate faba bean on raised beds nationally, faba bean gap will exist only in Middle Egypt and Lower and Upper Egypt will develop self sufficiency and can help in filling the gap in Middle Egypt. Thus, Egypt can have self sufficiency in faba bean and can have extra 262,096 tons (Table 6.12).

Furthermore, if we assumed that all national faba bean cultivated area will be irrigated with drip system and the intercropping will implemented, food gap will be exist in Middle Egypt only, i.e. 23,446 tones and it can be filled by the extra production of Upper Egypt. Thus, we can develop self sufficiency in faba bean and have extra 277,824 tons of faba bean seeds (Table 6.13).

Table 6.13 Faba bean total production and its gap under intercropping and drip system

	Faba bean total production (ton)	Faba bean food gap (ton)	Amount of extra faba bean (ton)
Lower Egypt	439,137	+111,077	111,077
Middle Egypt	92,950	−23,446	0
Upper Egypt	281,438	+190,193	166,747
Total	813,525		277,824

Effect of Climate Change on Faba Bean Production

Two consequences of climate change on crops are expected, namely drought and heat stress. Drought is an important environmental factor, which induces significant alterations in plant physiology and biochemistry. Faba bean is more sensitive to drought than some other seed legumes including common bean, pea and chickpea (Amede and Schubert 2003). Abd El-Mawgoud (2006) stated that the vegetative growth parameters, as well as yield components responded negatively to reduction in irrigation supply, namely plant height, number of leaves and fresh and dry weights. Faba bean growth stages responded differently to drought. The phase of pod set was found to be more sensitive to drought, followed by pod filling and the vegetative phase (Younise 2002). Furthermore, in faba bean, water stress decreases the final leaf area (Saxena 1991), net photosynthesis (Hura et al. 2007), light use efficiency (Xia 1994), pod retention and filling by reducing the availability of assimilate and distorting hormonal balances (Manschadi et al. 1998). However, mild water stress during flowering followed by sufficient irrigation of water after flowering resulted in slightly increased in seed yield and harvest index (Younise 2002).

Heat stress is another expected consequence of climate change. Vulnerability of faba bean to heat stress negatively affect floral stages, where flowers are the most affected during initial green-bud stages. Yield and pollen germination of flowers present before heat stress showed threshold relationships to stress, with lethal temperatures between 28 and 32 °C, whereas whole plant yield showed a linear negative relationship to stress with high plasticity in yield allocation, such that yield lost at lower nodes was partially compensated at higher nodal positions (Bishop 2016). These findings suggest that yield of faba bean will be limited by projected climate change, necessitating the development of drought and heat tolerant cultivars, or improved resilience by other mechanisms, such as earlier flowering times to escape drought and heat stresses periods.

Assessment of Faba Bean Production-Consumption Gap in 2030

To be optimistic, several assumptions were used in the assessment of faba bean production-consumption gap in 2030. The first assumption is the success of the breeding programs in producing faba bean cultivars with high tolerance to drought

Table 6.14 Projected faba bean cultivated area and production under climate change effect in 2029/2030

	Old land		New land		Total	
	Area (ha)	Production (ton)	Area (ha)	Production (ton)	Area (ha)	Production (ton)
Lower Egypt	23,708	81,683	12,633	48,257	36,341	129,940
Middle Egypt	1140	2787	23	63	1162	2851
Upper Egypt	2627	7974	1095	2826	3722	10,799
Total	27,475	92,444	13,750	51,145	41,225	143,590

Table 6.15 Projected Egyptian population, faba bean consumption and gap, new cultivated area to fill the gap and total required irrigation water in 2029/2030

	Population in july 2030	Total faba bean consumption (ton)	Faba bean food gap (ton)	New area to fill the gap (ha)	Total required water (m³)
Lower Egypt	76,547,694	484,547	354,607	92,829	405,353,537
Middle Egypt	27,588,466	174,635	171,784	61,496	310,894,147
Upper Egypt	21,734,576	137,580	126,781	49,127	319,870,240
Total	125,870,736	796,762	653,172	203,451	1,036,117,924

and heat stress. This assumption is supported by the ongoing breeding programs implemented by The Agricultural Research Center in Egypt. Thus, no expected yield losses will occur in 2030. The second assumption indicated that the assigned irrigation water for faba bean will be fixed as a result of no effect of climate change on the Nile flow. The third assumption was the per inhabitant consumption of faba bean will not change in 2030.

Projected Faba Bean Gap Under Surface Irrigation in 2029/2030

As it was stated in Chap. 2, Egyptian population is projected to be 125,870,736 inhabitants in 2030. In the growing season of faba bean in 2029/2030 and under surface irrigation; its total cultivated area will be reduced to 27,475 and 13,750 ha in the old and new lands, respectively with total production of 143,590 tons (Table 6.14).

In 2030, total faba bean consumption could reach 796,762 tons, which make faba bean consumption-production gap 653,172 tons. To fill this gap, 203,451 ha will need to be cultivated and that will require the application of 1,036,117,924 m³ (Table 6.15).

Thus, in 2030, if we continue using surface irrigation to grow faba bean its consumption-production gap will enlarged. Therefore, cultivation on raised beds or using drip system for faba bean irrigation, in addition to intercropping with other crops must be implemented to reduce that gap.

Projected Faba Bean Gap Under Cultivation on Raised Beds in 2029/2030

Cultivation of faba bean on raised beds can reduce the applied water by 20 % and increase productivity by 15 %, compare to irrigation with surface irrigation. Thus, the saved irrigation water can be used to cultivate a total of 8242 ha, which produce 30,158 tons of faba bean and that will increase total faba bean production to 173,748 tons (Table 6.16).

Changing faba bean cultivation to raised beds can reduce its gap to 623,014 tons. To fill this gap, cultivation of 195,209 ha on the national level is needed with total required water equal to 998,200,513 m^3 (Table 6.17).

Projected Faba Bean Gap Using Drip System in 2029/2030

Irrigation of faba bean with drip system could increase the production of old land to be 101,689 tons and allow to cultivate extra 13,737 ha with faba bean using the saved irrigation water as a result of increasing application efficiency. This procedure will increase faba bean national production to be 203,098 tons (Table 6.18).

Furthermore, this procedure will result in reduction of faba bean production-consumption gap to be 593,664 tons. Cultivation of 187,167 ha with the application of 961,068,729 m^3 of irrigation water will be required to fill that gap (Table 6.19).

Table 6.16 Projected new cultivated area with faba bean, its production and total production under raised beds cultivation in Egypt in 2029/2030

	New cultivated area (ha)	Production of new area (ton)	Total cultivated area (ha)	Total production (ton)
Lower Egypt	7112	27,169	43,453	157,109
Middle Egypt	342	955	1504	3806
Upper Egypt	788	2034	4510	12,833
Total	8242	30,158	49,467	173,748

Table 6.17 Projected faba bean food gap, new cultivated area to fill the gap and total required irrigation water under raised beds cultivation in 2029/2030

	Faba bean food gap (ton)	New area to fill the gap (ha)	Total required water (m^3)
Lower Egypt	327,438	85,717	374,296,106
Middle Egypt	170,829	61,154	309,165,887
Upper Egypt	124,747	48,339	314,738,520
Total	623,014	195,209	998,200,513

Table 6.18 Potential faba bean production under drip irrigation in the old and new lands of Egypt in 2029/2030

	Production of old land (ton)	New cultivated area (ha)	Production of new area (ton)	Total cultivated area (ha)	Total production (ton)
Lower Egypt	89,852	11,854	45,282	48,195	183,390
Middle Egypt	3066	570	1592	1732	4721
Upper Egypt	8771	1314	3390	5036	14,987
Total	101,689	13,737	50,264	54,962	203,098

Table 6.19 Faba bean food gap, new cultivated area to fill the gap and total required irrigation water under drip system in 2029/2030

	Faba bean food gap (ton)	New area to fill the gap (ha)	Total required water (m³)
Lower Egypt	301,156	78,837	344,253,868
Middle Egypt	169,914	60,826	307,509,247
Upper Egypt	122,593	47,504	309,305,613
Total	593,664	187,167	961,068,729

Table 6.20 Projected faba bean production under different systems of intercropping in 2029/2030

	Faba bean potential production (ton) under intercropping with				
	Tomato	Sugar beet	Sugarcane	Fruit trees	Total (ton)
Lower Egypt	34,103	59,112	Not cultivated	130,844	224,059
Middle Egypt	25,999	11,860	Not cultivated	13,432	51,291
Upper Egypt	19,477	1443	50,560	20,271	91,751
Total	79,579	72,416	50,560	164,546	367,102

Faba Bean Intercropping with Other Crops

Intercropping faba bean with the suggested crops will result in an increase in faba bean total production to be 367,102 tons (Table 6.20).

Thus, added faba bean total production from intercropping to faba bean production resulted from cultivation on raised beds nationally; it will reduce the gap to 237,629 tons. To fill this gap, we need to cultivate 76,520 ha with faba bean under drip system. Furthermore, the total required water to cultivate this area will be 384,106,972 m³ (Table 6.21).

Furthermore, if we assumed that all national faba bean cultivated area will be irrigated with drip and we will implement intercropping, the gap will be reduced to be 208,279 tons. A total new cultivated area of 68,601 ha is need to be cultivated to produce faba bean seeds to fill the gap, which will require 347,907,030 m³ of irrigation water (Table 6.22).

Table 6.21 Projected faba bean food gap, new cultivated area to fill the gap and total required irrigation water under intercropping and raised beds cultivation in 2029/2030

	Faba bean total production (ton)	Faba bean food gap (ton)	New area to fill the gap (ha)	Total required water (m^3)
Lower Egypt	395,707	88,840	23,441	102,359,103
Middle Egypt	56,589	118,046	43,871	221,790,809
Upper Egypt	106,837	30,743	9208	59,957,060
Total	559,133	237,629	76,520	384,106,972

Table 6.22 Faba bean food gap, new cultivated area to fill the gap and total required irrigation water under intercropping and sprinkler system in 2029/2030

	Faba bean total production (ton)	Faba bean food gap (ton)	New area to fill the gap (ha)	Total required water (m^3)
Lower Egypt	421,988	62,559	16,507	72,078,585
Middle Egypt	57,504	117,131	43,531	220,070,956
Upper Egypt	108,990	28,590	8563	55,757,489
Total	588,483	208,279	68,601	347,907,030

Conclusion

Faba bean is one of the most important leguminous crops for human nutrition, is also strategic due to its income contribution in Egypt, as well as being significant for soil fertility. The current faba bean production-consumption gap is 73 %. This gap can be reduced to 64 %, if we change cultivation method in the old lands from basins or narrow furrows to raised beds. More reduction in faba bean gap can be obtained if we irrigate all the cultivated area of faba bean with drip system, where the gap will be 61 %. Self sufficiency in faba bean can be attained if we implement the suggested intercropping systems and cultivation on raised beds or if we implement the suggested intercropping systems and we use drip system for irrigation in the old land. Thus, an extra amount of faba bean can be obtained, namely 262,096 or 277,824 tons of faba bean seeds, if we implemented one of these techniques, respectively.

However, under climate change in 2030 and surface irrigation, the gap will be 82 %. Furthermore, the gap will be 78 %, if cultivation of faba bean in the old cultivated land on raised beds will be implemented. Irrigation of faba bean with drip system in the old land will reduce the gap to 75 %. Implement of cultivation on raised beds and intercropping systems can reduce the gap to 30 % and if we irrigate faba bean in the old land with drip system and implement intercropping systems, the gap will be 26 %.

References

Abd El-Zaher Sh, R., & Gendy, E. K. (2014). Effect of plant density and mineral and bio-nitrogen fertilization on intercropping faba bean with sugar beet. *Egyptian Journal of Applied Sciences, 29*(7), 352–366.

Abdel-Mawgoud, A. M. R. (2006). Growth, yield and quality of green bean (Phaseolus vulgaris) in response to irrigation and compost applications. *Journal of Applied Sciences, 2*(7), 443–450.

Abouelenein, R., Oweis, T., El Sherif, M., Awad, H., Foaad, F., Abd El Hafez, S., Hammam, A., Karajeh, F., Karo, M., & Linda, A. (2009). Improving wheat water productivity under different methods of irrigation management and nitrogen fertilizer rates. *Egyptian Journal of Applied Sciences, 24*(12A), 417–431.

Addo-Quaye, A. A., Darkwa, A. A., & Ocloo, G. K. (2011). Growth analysis of component crops in a maize-soybean intercropping system as affected by time of planting and spatial arrangement. *ARPN Journal of Agricultural and Biological Science, 6*, 34–44.

Al-Suhaibani, N., El-Hendawy, S., & Schmidhalter, U. (2013). Influence of varied plant density on growth and economic return of drip irrigated faba bean. *Turkish Journal of Field Crops, 18*(2), 185–197.

Amede, T., & Schubert, S. (2003). Mechanisms of drought resistance in seed legumes. I. Osmotic adjustment. *Ethiopian Journal of Science, 26*, 37–46.

Bishop, J. (2016). Responses of faba bean (Vicia faba L.) yield parameters to heat stress treatments for five days during floral development and anthesis. University of Reading. http://dx.doi.org/10.17864/1947.48

Dean, G., & Mendham, N. (2003). Optimum plant densities for faba bean cv Fiesta VF sown on raised beds. Australian Society of Agronomy."Solutions for a better environment". Proceedings of the 11th Australian Agronomy Conference Geelong, Victoria.

Farghly, B. S. (1997). Yield of sugar cane as affected by intercropping with faba bean. *The Journal of Agricultural Science, Mansoura University, 22*(12), 4177–4186.

Fischer, R. A., Sayre K., & Ortiz Monasterio, I. (2005). The effect of raised bed planting on irrigated wheat yield as influenced by variety and row spacing. In: *Evaluation and performance of permanent raised bed cropping systems in Asia, Australia and Mexico.* ACIAR Proceedings No. 121.

Hura, T., Hura, K., Grzesiak, M., & Rzepka, A. (2007). Effect of long term drought stress on leaf gas exchange and fluorescence parameters in C3 and C4 plants. *Acta Physiologiae Plantarum, 29*, 103–113.

Ibrahim, S. (2006). Intercropping potato with maize. *Annals of Agricultural Science, 44*(3), 799–817.

Kumar, A., Sharma, K. D., & Yadav, A. (2010). Enhancing yield and water productivity of wheat (Triticum aestivum) through furrow irrigated raised bed system in the Indo Gangetic Plains of India. *Indian Journal of Agricultural Sciences, 80*(3), 198–202.

Li, L., Sun, J. H., Zhang, F. S., Li, X. L., & Rengel, Z. (2001). Wheat/maize or wheat/soybean strip intercropping: I. Recovery or compensation of maize and soybean after wheat harvesting. *Field Crops Research, 7*, 173–181.

Li, L., Li, S. M., Sun, J. H., Zhou, L. L., Bao, X. G., Zhang, H. G., & Zhang, F. S. (2007). Diversity enhances agricultural productivity via rhizosphere phosphorus facilitation on phosphorus-deficient soils. *Proceedings of the National Academy of Sciences of the United States of America, 104*(27), 11192–11196.

Limon-Ortega, A., Sayre, K. D., Drijber, R. A., & Francis, C. A. (2002). Soil attributes in a furrow-irrigated bed planting system in northwest Mexico. *Soil Tillage Research, 63*, 123–132.

Manschadi, A. M., Sauerborn, J., Stutzel, H., Gobel, W., & Saxena, M. C. (1998). Simulation of faba bean (Vicia faba L.) growth and development under Mediterranean conditions: Model adaptation and evaluation. *European Journal of Agronomy, 9*, 273–293.

Mansour, H. A., Sabreen, K., Pibars, M., El-Hady, A., & Eldardiry, E. I. (2014). Effect of water management by drip irrigation automation controller system on faba bean production under water deficit. *International Journal of Geomate, 7*(2), 1047–1053.

Nazir, M. S., Jabbar, A., Ahmad, I., Nawaz, S., & Bhatti, I. H. (2002). Production potential and economics of intercropping in autumn-planted sugarcane. *International Journal of Agriculture and Biology, 4*(1), 140–141.

Noreldin, T., Ouda, S., & Taha, A. (2015). Combating adverse consequences of climate change on maize crop. In *Major crops and water scarcity in Egypt* (pp. 53–67). Springer Publishing House.

Poutney, N. (2014). Faba bean marketing and the Middle East. Grain Research and Development Corporation in Australia https://grdc.com.au/Research-and-Development/GRDC-Update-Papers/2014/07/Faba-bean-marketing-and-the-Middle-East

Pramanik, S. C., Singh, N. B., & Singh, K. K. (2009). Yield, economics and water use efficiency of chickpea (Cicerarietinum) under various irrigation regimes on raised bed planting system. *Indian Journal of Agronomy, 54*(3), 315–318.

Raghothama, K., & Karthikeyan, A. (2005). Phosphate acquisition. *Plant and Soil, 274*, 37–49.

Saxena, M. C. (1991). Status and scope for production of faba bean in Mediterranean countries. *Options Méditerranéennes, 10*,1 5-20-47.

Singh, A. K., & Umrao, V. K. (2013). GAP: Monetary way to manage faba bean diseases: A review. *HortFlora Research Spectrum, 2*(2), 93 102.

Singh, A. K., Bhatt, B. P., Sundaram, P. K., Gupta, A. K., & Singh, D. (2013). Planting geometry to optimize growth and productivity in faba bean (Vicia faba L.) and soil fertility. *Journal of Environmental Biology, 34*(1), 117–122.

Snyder R. L., Orang M., Bali K., & Eching S. (2004).Basic irrigation scheduling BIS. http://www.waterplan.water.ca.gov/landwateruse/wateruse/Ag/CUP/Californi/Climate_Data_010804.xls

WOCAT. (2016). Mixed fruit tree orchard with intercropping of Esparcet and annual crops in Muminabad District Tajikistan: Bog Orchard based agroforestry established on the hill slopes of Muminabad. WOCAT_QT_Summary-T_TAJ043en.pdf.

Xia, M. Z. (1994). Effects of soil drought during the generative development phase of faba bean (Vicia faba L.) on photosynthetic characters and biomass production. *Journal of Agricultural Science, 122*, 67–72.

Younise, I. D. (2002). Effect of water stress at different growth stages on growth and productivity of faba bean (Vicia faba L.). M.Sc. thesis, University of Khartoum.

Chapter 7
Conclusion and Recommendations to Policy Makers

Samiha A.H. Ouda, Abd El-Hafeez Zohry, and Ahmed Said Kamel

Introduction

Agriculture in Egypt is considered one of the most important sources of life; it con-tributes in achieving comprehensive and sustainable development of the society. The growing importance of agriculture for the time being is due to the large food gaps, which influences the national economy of the major cereal crops, and makes the problem of food security one of the most important priorities that need to be addressed (El-Sadek 2010). Food gaps problems are occupying great importance in the light of the steady increase in population, so as the demand on major food com-modities, and lack of resources (Hafez et al. 2011). A wide gap in self-sufficiency between the production and consumption of food occurred in the light of the increas-ing population, rising standards of living, declining of trade in food grains and high prices in the market (Gerber 2014).

The concept of food security is based on three main pillars, food availability, food accessibility and food stability. We were concerned in this book with food availability represented by wheat, maize and faba bean as major food commodities in Egypt. The Egyptian government has attempted over centuries to provide food for the Egyptian people and achieve food security, including policies to increase domestic production (WFP 2013). This is done alongside of adopting economic and social development plans to support agricultural activities through expansion of agriculture lands, improvement of seeds, technologies, irrigation water facilities, roads and services (Lewis 2011). However, high population growth rate defeated all

S.A.H. Ouda (✉)
Water Requirements and Field Irrigation Research Department, Soils, Water and Environment Research Institute, Agricultural Research Center, Giza, Egypt
e-mail: samihaouda@yahoo.com

A.E.-H. Zohry • A.S. Kamel
Crops Intensifications Research Department, Field Crops Research Institute,
Crops Agricultural Research Center, Giza, Egypt

© Springer International Publishing AG 2017
S. Ouda et al., *Future of Food Gaps in Egypt*, SpringerBriefs in Agriculture,
DOI 10.1007/978-3-319-46942-3_7

these attempts. Therefore, assessment of unconventional procedures need to done to reduce these food gaps in the light of its expected increase in the future due to reduction in the cultivated area of these crops.

Lots of scattered studies were done in Egypt on increasing productivity of wheat, maize and faba bean on farm level. These studies were either on using better water management on farm level, using improved fertilizers application and/or pests and weed control. Land system for food production is considered the main factor for achieving food security. It includes the availability of land for agriculture and the management of land for stable food production (Verburg et al. 2013). Although Egypt is blessed by having lands suitable for cultivation, it suffers from limited water resources, which require better management for lands and water resources.

In this book, we assessed different options to increase the production of these three important crops, where population increase and climate change effect pose as huge obstacles. Thus, in this chapter, we presented the important results obtained in Chaps. 4, 5 and 6 to come with recommendations to policy makers.

Wheat Current and Future Production in 2030

The analysis in Chap. 4 showed that wheat production-consumption gap in 2012/2013 growing season was 49 %. This means that the national wheat production contributed with 51 % of the total wheat consumption in Egypt. Under these circumstances, wheat is grown in basins under surface irrigation with 60 % application efficiency. Changing cultivation methods to raised beds can save 20 % of the applied irrigation water and increase wheat yield by 15 % (Abouelenein et al. 2009). Thus, wheat production under raised beds can attain 68 % of national wheat consumption, with 17 % increase, compare to basins cultivation. Irrigating wheat with sprinkler on national level can increase the contribution of national production to national consumption to 72 %, with 21 % increase, compare to basins cultivation. Previous research on the effect of using sprinkler system to irrigate wheat revealed that it can save 20 % of the applied irrigation water and yield can increase by 18 % (Taha 2012). The highest contribution to national consumption can be attain if we irrigate the assigned area to cultivate wheat with sprinkler system and implement the suggested intercropping systems (wheat intercropping on sugar beet, tomato, cotton, sugarcane and under young fruit trees). This procedure can produce high wheat yield and contribute to national consumption by 81 %, with 30 % increase, compare to basins cultivation (Table 7.1). Previous research on intercropping wheat with the suggested crops revealed that wheat plants are grown on the lands originally assigned to these crops, where wheat shared its lands and its applied irrigation water (Ouda and Zohry 2016).

In 2029/2030 wheat growing season, where climate change effect will be pronounced, the cultivated area of wheat will be reduced as a result of increase wheat water requirements and fixed amount of irrigation water allocated to wheat. Under these circumstances, wheat production-consumption gap will be 62 % and wheat

Table 7.1 Percentage of contribution of wheat production in the total consumption in 2012/2013 and 2029/2030 growing seasons

	2012/2013 season	Percentage of increase (%)	2029/2030 season	Percentage of increase (%)
Basins	51	–	37	–
Raised beds	68	17	45	8
Sprinkler system	72	21	56	19
Raised beds + intercropping	77	26	51	14
Sprinkler + intercropping	81	30	62	25

national production will contribute by only 37 % in the national consumption. This situation will occur if we continue to use basins for wheat cultivation. Application of either one of the options exist in Table 7.1 will result in increase the percentage of the contribution of national wheat consumption, where the highest contribution will occur if we used sprinkler system for all wheat cultivated area and we implement intercropping systems. In this case, percentage of the contribution will be 62 %, with 25 % increase, compare to basins cultivation (Table 7.1).

It worth noting that all the options existed in Table 7.1 to increase wheat national production did not use any extra irrigation water; instead it used the assigned water to irrigate wheat with more rationale use and that can increase wheat production and contribution to nation consumption, thus it can reduce wheat importation.

Maize Current and Future Production in 2030

In 2013 maize growing season, the production-consumption gap was 47 %. In that season maize was grown in basins under surface irrigation with 60 % application efficiency. If we cultivate maize on raised beds, its national production can increase, so as its percentage to national consumption by 20 % from 53 % under basins cultivation to 73 % under raised beds cultivation. Furthermore, 30 % increase in the national maize production can be attained if we change its irrigation system from surface to drip. Drip system has proved to save 30 % of the applied irrigation water and increase yield by 18 % (Taha 2012). In this case, maize production can contribute by 83 % to its national consumption. Furthermore, implementing maize intercropping systems with other crops (tomato, potato, peanut, soybean and cowpea) and irrigate all the assigned area for maize with drip system will result in high contribution to maize national consumption, namely 86 %, compared to maize cultivation in basins (Table 7.2).

The contributions of the options exist in Table 7.2 in maize national consumption will be lower in 2030 under climate change effect. The lowest contribution will occur if we use basins and the highest contribution will occur if we use drip system and intercropping, i.e. 47 % versus 69 %, respectively.

Table 7.2 Percentage of contribution of maize production in the total consumption in 2013 and 2030 growing seasons

	2013 season	Percentage of increase (%)	2030 season	Percentage of increase (%)
Basins	53	–	47	–
Raised beds	73	20	59	12
Drip system	83	30	66	19
Raised beds + intercropping	76	23	62	15
Drip + intercropping	86	33	69	22

Table 7.3 Percentage of contribution of faba bean production in the total consumption in 2012/2013 and 2029/2030 growing seasons

	2012/2013 season	Percentage of increase (%)	2029/2030 season	Percentage of increase (%)
Basins	27	–	18	–
Raised beds	36	8	22	4
Drip system	39	11	25	7
Raised beds + intercropping	149	122	70	52
Drip + intercropping	152	125	74	56

Furthermore, all the options existed in Table 7.2 does not require any extra irrigation water; instead it used the assigned water to irrigate maize more rationally, which resulted in an increase in national maize production, thus reduce maize importation.

Faba Bean Current and Future Production in 2030

With respect to faba bean production-consumption gap, Chap. 5 revealed that it is 73 % and we only produce 27 % of our consumption. Table 7.3 revealed that cultivation of faba bean on raised beds and implementing intercropping systems (faba bean intercropped with sugar beet, tomato and sugarcane, as well as under young fruit trees) can increase its production by 122 %, compared to its production under basins cultivation. Furthermore, 125 % increase in faba bean production can be attained if we irrigate faba bean with drip system and implementing intercropping systems (Table 7.3). The results in Chap. 6 revealed that faba bean intercropping systems can add 607,185 tons of faba bean seeds, whereas our national consumption is 535,701 tones of faba bean seeds. This result implied that we should cultivate faba bean only in intercropping systems and designate its assigned cultivated area to be cultivated with other winter crops, like wheat which could minimize its production-consumption gap.

Unfortunately, under climate change in 2029/2030 growing season, faba bean intercropping systems will have lower contribution to faba bean national consumption (Table 7.3). This could be attributed to the increase in faba bean water require-ments and loss in its cultivated area, which will highly reduce faba bean national production and increase its production-consumption gap.

Conclusion

Increasing production of wheat, maize and faba bean can be done by increasing the cultivated area, and emphasizing the importance of agricultural investment to achieve higher production efficiency and it can be done by technical cooperation, transfer of expertise, agricultural research, providing efficient human resources and updated technologies to achieve the development of sustainable agricultural produc-tivity and food security. However, limitation of water resources will be an obstacle to achieve that. Therefore, improving irrigation water management can use water more rationally and help in cultivating more lands with the same amount of assigned water for agriculture. Availability of irrigation water could help in the transforma-tion of agricultural land from non-cultivated to cultivated lands. Furthermore, inter-cropping of these three important crops with other crops can increase its production and reduce its gap.

Importation of wheat, maize and faba bean to covers the gap between the domes-tic supply and population needs costing the government large amounts of Egyptian pounds. Thus, there is a need for a political and an economic stability factor of the local market to maintain sustainable development, by managing the factors of price increase and limiting the imports.

References

Abouelenein, R., Oweis, T., El Sherif, M., Awad, H., Foaad, F., Abd El Hafez, S., et al. (2009). Improving wheat water productivity under different methods of irrigation management and nitrogen fertilizer rates. *Egyptian Journal of Applied Science, 24*(12A), 417–431.

El-Sadek, A. (2010). Virtual water trade as a solution for water scarcity in Egypt. *Water Resource Management, 24,* 2437–2448.

Gerber, A. (2014). *Food security as an outcome of food systems.* http://www.systemdynamics.org/conferences/2014/proceed/papers/P1113.pdf

Hafez, W. et al. (2011). *Food security in Egypt in 2030: Future scenarios.* http://www.idsc.gov.eg/IDSC/publication/View.aspx?ID=352

Lewis, L. N. (2011). *Egypt's future depends on agriculture and wisdom.* http://aic.ucdavis.edu/calmed/August%2008,%20final%202d%20Egypt.pdf

Ouda, S. & Zohry, A. A. (2016). Significance of reduction of applied irrigation water to wheat crop. In *Major crops and water scarcity in Egypt* (pp. 33–50). Springer Publishing House.

Taha, A. (2012). *Effect of climate change on maize and wheat grown under fertigation treatments in newly reclaimed soil.* PhD thesis, Tanta University, Egypt.

Verburg, P. H., et al. (2013). Land system change and food security: Towards multi-scale land system solutions. *Current Opinion in Environmental Sustainability, 5*(5), 494–502.

WFP. (2013). *The status of poverty and food security in Egypt: Analysis and policy recommenda-tions.* http://documents.wfp.org/stellent/groups/public/documents/ena/wfp257467.pdf

Printed in the United States
By Bookmasters